Mind in Nature

Mind in Nature

John Dewey, Cognitive Science, and a Naturalistic Philosophy for Living

Mark L. Johnson and Jay Schulkin

The MIT Press
Cambridge, Massachusetts
London, England

The MIT Press
Massachusetts Institute of Technology
77 Massachusetts Avenue, Cambridge, MA 02139
mitpress.mit.edu

The MIT Press would like to thank the anonymous peer reviewers who provided comments on drafts of this book. The generous work of academic experts is essential for establishing the authority and quality of our publications. We acknowledge with gratitude the contributions of these otherwise uncredited readers.

This book was set in Stone Serif by Westchester Publishing Services.

Library of Congress Cataloging-in-Publication Data

Names: Johnson, Mark, 1949– author. | Schulkin, Jay, author.
Title: Mind in nature : Dewey, cognitive science, and a naturalistic philosophy for living / Mark L. Johnson and Jay Schulkin.
Description: Cambridge, Massachusetts : The MIT Press, [2023] | Includes bibliographical references and index.
Identifiers: LCCN 2022017648 (print) | LCCN 2022017649 (ebook) | ISBN 9780262545167 | ISBN 9780262373456 (epub) | ISBN 9780262373463 (pdf)
Subjects: LCSH: Philosophy of mind. | Philosophy and cognitive science. | Dewey, John, 1859–1952. | Dewey, John, 1859–1952. Experience and nature.
Classification: LCC BD418.3 .J645 2023 (print) | LCC BD418.3 (ebook) | DDC 128/.2—dc23/eng/20220919
LC record available at https://lccn.loc.gov/2022017648
LC ebook record available at https://lccn.loc.gov/2022017649

150377129

From Mark, to Scott Pratt,
who has shown me the marvelous breadth
and rich diversity of American Philosophy

From Jay, to Ernie Hook, son of Sidney Hook,
and to Tibor Solymosi, wonderful colleagues

Contents

Preface

This has been a labor of great joy. The two of us have known each other for about ten years. We liked each other's work, learned from each other, and discovered our shared interest in the possibilities for a fruitful collaboration between pragmatist philosophy and recent cognitive science. We published an article on Dewey's idea that our aesthetic sensibility is the common currency of human experience, meaning, and thought. Jay was just about to begin a new book on the wonders of Dewey's great book *Experience and Nature*, and Mark asked if he could tag along on the project. We were both enamored of this amazing treatise. What followed, during the pandemic, was a delightful series of weekly Zoom meetings in which we worked through every chapter of *Experience and Nature*, taking as much time as we needed on any point that we were puzzled about, or wanted to explore in more depth.

The contours of this present book began to emerge from our weekly discussions. We saw Dewey's pragmatism as providing an excellent overarching framework for making sense of a growing body of contemporary scientific research on experience, meaning, mind, thought, consciousness, selfhood, and values. Jay's neuroscience background supplied empirical research that allowed us to expand, enrich, and, occasionally, criticize Dewey's major claims. Along our journey, we came to further appreciate Dewey's naturalistic philosophy for living. It is a philosophy that is non-dualistic, rooted in our embodiment, shaped by experience, and based on fundamental biological and social values.

Our many conversations were such fun for the two of us as we puzzled over the text. What were Dewey's most profound insights about human nature? What does a naturalized philosophy reveal about mind, meaning, and thought? Where was Dewey clear and where not? Even when he wasn't clear, could we grasp what he was trying to say, and could we carry his thought forward in a way that meshed with cognitive science and neuroscience? Above all, Dewey had a profound grasp of what mattered, or ought to matter, for our lives. He constantly reminds us to pay close attention to the depths of experience, and to cultivate an appropriate critical perspective on the values driving our inquiries and moral deliberations. The result is a naturalistic perspective rich enough to explain how the processes of our biological and social viability shape our experience, meaning, thought, and values, with a clear linkage to the biological, cognitive, and neurosciences that provide a basis for an updated, expanded, and existentially meaningful version of an *Experience and Nature* suited to the twenty-first century.

The result is this book.

Mark L. Johnson and Jay Schulkin

June 2022

1 Introduction: Philosophy, Naturally

Toward a Natural(istic) Philosophy to Live By

In 1925 John Dewey published *Experience and Nature,* one of the most important philosophical works ever written. This sweeping assertion may sound arrogant or perhaps even delusional—completely out of touch with reality. After all, outside of academic philosophy, in the world of generally educated people, who has ever heard of John Dewey, let alone of this particular book? No, he's not the guy who created the Dewey Decimal System, and he's not the Dewey who lost the 1948 presidential race to Truman. He is John Dewey, the American public intellectual who had a substantial influence on philosophy, and especially on educational theory, in the first half of the twentieth century, but whose work fell out of favor and was eclipsed by the linguistic analytic philosophy that arose in the 1930s and still flourishes today. The logical empiricist predilection for linguistic analysis and logical rigor pushed pragmatism in general into the shadows, except in the context of the more scientific dimensions of pragmatism (Godfrey-Smith 1998; Kitcher 2012, 2018; Quine 1951, 1953/1961, 1960). In analytic philosophy, Dewey is typically viewed as an unclear, nonrigorous thinker whose prose is obscure, turgid, and ambiguous. In the European traditions of phenomenology and hermeneutics, Dewey is generally regarded as overly scientistic and not existentially engaged. To the contrary, we will argue that Dewey is a model of rigor and precision, where those standards are appropriate, and he is a stellar model of engagement with the affairs of everyday life, arguing that philosophy should arise in response to problems encountered in everyday experience and then reconstruct problematic experiences to make things better. So, we will see that both of these traditional takes on

Dewey's philosophy are fundamentally mistaken, even though one can see how these standard misreadings arise.

The eclipse of American pragmatism by linguistically oriented analytic philosophy lasted well into the 1970s, but the past four decades have witnessed a remarkable renaissance of pragmatist philosophy. This rediscovery of pragmatism began as the result of a heated debate within the American Philosophical Association (APA) concerning what type of philosophy should be presented at annual meetings of the three divisions of the APA. Basically, it was a debate about what philosophy should be and do. In a contentious vote, members familiar with classical American (and also continental) philosophy were admitted to committees responsible for selecting papers for presentation at national meetings of the APA. Some of the stalwarts of American pragmatism—John McDermott, Elizabeth Flower, John Smith, Murray Murphy, Abraham Edel, Philip Wiener, Max Fisch, Sidney Hook, Morton White, and several others—had kept some classical pragmatist writings in print during the years when pragmatism fell out of favor. Then, with the renewed interest in pragmatism in both the APA and in the newly formed Society for the Advancement of American Philosophy in the early 1970s there was a rebirth of interest in classical American philosophy (see appendix).

Against this general background of growing recognition of the exciting new contributions of pragmatist thought to multiple domains of human experience, our main thesis is threefold: (1) that in *Experience and Nature* Dewey presents the most important and compelling naturalistic philosophy ever penned, (2) that a good deal of contemporary science and philosophy supports and enriches Dewey's philosophical perspective, thereby confirming our generous assessment of his work, and (3) that Dewey gives us a profound philosophy to live by. Tracking these three themes, most of the following chapters begin with an interpretation of some of Dewey's most radical views on the nature of experience, mind, thought, selfhood, knowledge, and values. Along the way, we bring to bear contemporary scientific research from biology, neuroscience, psychology, and cognitive science, in order to elaborate, deepen, and sometimes criticize Dewey's central claims. For the most part, we find that relevant scientific research confirms many of Dewey's most profound insights. Finally, we are concerned to work out the implications of this naturalistic philosophy for how we ought to live, and what kind of person we should strive to become.

By a naturalistic philosophy, we mean the view that everything that exists is part of nature, which thereby encompasses all of the natural processes by which things and events come into, persist for a time through, and pass out of existence. A naturalistic philosophy employs empirical methods of inquiry and explanation that have no need to postulate supernatural or transcendent agents, entities, causes, or forces. From a naturalistic perspective, therefore, humans are complex, highly evolved biological and social animals engaging their environments. We are always in and of nature, and we are a pivot, however small, of how some portion of experience is currently unfolding and developing. Our embodied, social animality is not a fallen condition to be endured and lamented; rather, it is the ground of all our experience and meaning, and it is the source of our most glorious achievements.

Some folks will no doubt think that any such naturalism must be, if not nihilistic, at least deflationary about human capacities, dignity, exceptionalism, values, and possibilities for meaningful existence. They ask: Are we merely fancy animals mechanistically driven to perceive, value, and act in habitually established paths of behavior, working out our fragile existence over our short lifespan, only to eventually dissolve into the earth from which we originally emerged?

Our answer to that fundamental question concerning the human condition will be both "Yes" and "No." The answer is yes, insofar as we are just sophisticated animals who, as Macbeth feared, strut and fret our hour upon the stage, thence to be heard no more. But, the answer is no, insofar as the right kind of naturalism recognizes our capacity to live a meaningful life, a life with value and purpose, yet realized wholly within the natural world. We will argue that a naturalistic approach of this sort is not entirely reductionistic; indeed, it provides a rich account of the development of our human capacities to experience, act, think, create, and value, and it can offer realistic guidance for how we might live a purposeful and morally intelligent life.

Why Dewey?

Why focus primarily on just one philosopher when there are many other respected pragmatist philosophers who have written eloquently and insightfully about many of the issues we take up here? For example, in previous books (Crippen & Schulkin 2020; Moreno & Schulkin 2020; Schulkin 2000, 2004, 2009, 2012) one of us has described the prescient insights of some of the early classical pragmatists like Peirce, James, Dewey, and Mead, who anticipated later discoveries in science and philosophy. These four understood the profound importance of action in all human perception, valuing, and thinking. With acknowledgment of the significant contributions of these early pragmatists, we have put Dewey center stage, for the following reasons: First, he is the only pragmatist to write about nearly every field of philosophy (e.g., metaphysics, epistemology, philosophy of mind, logic, value theory, ethics, politics, law, aesthetics, education, and social psychology) and to see fundamental interconnections among all of these fields. Dewey ranks with Aristotle for the remarkable scope of his knowledge of philosophy, the various sciences, the humanities, and the arts. Second, Dewey stands out for his grasp of the biological processes responsible for action, as the basis for all "higher" cognition. He anticipates contemporary pragmatists, such as Madzia and Jung (2016), Solymosi and Shook (2014), and Schulkin (2012), who see action as the key to our perception, motive control, and thought. He saw action as a matter of the adaptive adjustment of an organism to its environment and to other people (Engel et al. 2015; Friston et al. 2016; Peters et al. 2017). Dewey presaged the recognition that actions are shaped in part by the need to conserve energy in neural systems (Friston 2010; Sterling 2020), and also by the need to sustain predictive regulation within the animal in relation to its environment. Third, Dewey tenaciously diluted diverse forms of dualism, such as thought versus action and perception versus thought, arguing that "there is simply a continuously ordered sequence of events" (Dewey 1896, 364) via ongoing integration of afferent and efferent processes.

In short, we see Dewey as the best proponent of this view of motivated prediction-based action within a physical and socio-cultural environment. *Experience and Nature* is the fullest presentation of this perspective. C. S. Peirce, outstanding in so many ways, still wanted to keep logic pure, untainted by facts of transactions with others. James, who is so profoundly insightful about human attention, memory, experience, perception, and habit formation, never gave us a good sense of individuals co-evolving together with others—interacting, and transacting—at least not with the depth and scope we find in Dewey.

Dewey developed a philosophy of nature that gives us a sense of being at home in the world and a sense that our evolution is both natural and cultural, insofar as physical and social goings-on are endlessly permeable. Dewey captures insights later developed in Gestalt theory, and also in

Gibson's theory of affordances that shape our perception and action. There are clear parallels between Dewey and Merleau-Ponty in terms of their ontological, epistemological, and aesthetic commitments (Chemero 2009; Chemero & Kaufer 2016; Gallagher 2018), although Dewey's work seems more closely connected to biological accounts of the conditions of animal viability and the basis for adaptive prediction and action. In *Experience and Nature* Dewey is bent on action, presaging the profound implications of our embodiment for perception, attention, action, and values, in a way that is quite compatible with contemporary "enactivist" views (Gallagher 2005, 2016; Gallagher & Zahavi 2007; Thompson 2007; Varela et al. 1991).

Why Bother?

Why care about any of this? Our reply is twofold. First, we are generally better able to resolve problems, meet the challenges we face, and manage the contingencies of our existence, the more we cultivate our reflective intelligence, and the more we draw on the best empirical knowledge currently available concerning our self and our world. Second, anyone who feels the lure of the ancient injunction to "know thyself," will most likely appreciate Socrates' claim that the beginning of wisdom is to recognize one's own ignorance and to use that as a spur to the pursuit of knowledge of self and world. This requires an honest and impartial self-assessment of what it means to be human, that is, of both the limitations and possibilities for the kinds of creatures we are, or might become. Among other things, self-knowledge requires us to understand where our remarkable capacities come from, how they operate in our daily lives, and to what extent we have any control over what happens and who we are becoming. The underlying assumption is that reliable knowledge of self, others, and the world—knowledge that is empirically supported and interpersonally validated—is necessary if we want a nondelusional understanding upon which to stake our life values and projects. We are going to suggest that Dewey is almost without equal when it comes to developing such a naturalistic humanism.

The Structure of Our Argument

In order to appreciate the magnitude of Dewey's contribution to a meaningful and practical philosophical naturalism, it is necessary to understand the broad scope of his views about a number of important topics. Although a suitably comprehensive naturalistic theory would encompass inanimate nature, living nature (including plants and other animals besides ourselves), and mind, we will here focus primarily on the human condition, because it is about that condition that people are most existentially engaged, and from whence they tend to draw their views about the nonhuman world.

Dewey's project begins from the understanding that experience is the result of the ongoing interactions of an organism and its environments. Consequently, all of our capacities, thoughts, values, and actions arise as dimensions of organism–environment transactions, in a continuous fashion that avoids any ontological or epistemic dualisms. The task of such a nondualistic philosophical naturalism is to construct a suitably rich account of experience that explains how the increasing complexity of the organism, of its biological and sociocultural environments, and of their continuous interactions gives rise to a number of emergent functions, such as perception, movement, object manipulation, feeling, emotion, motivation, critical valuation, language, and reasoning. These emergent functional organizations ultimately issue in what we call mind and self. The resulting self is a joint product of evolution—which gives rise to the basic architecture of our bodies and brains—and the development of individual selves over their lifespan, at both the biological and sociocultural levels of organization.

This book is organized, therefore, around the increasing complexity of organisms and their environments that results in the characteristic emergent functions of human feeling, thought, action, and valuing. We begin with Dewey's account of how experience roots in the ongoing organism–environment patterns of transaction. The comprehensiveness, depth, and richness of one's initial description of experience fatefully shapes what they will count as important, and how they will understand and explain that experience. The primacy of experience cannot be overemphasized, insofar as it will determine what you count as relevant phenomena and what you count as an adequate explanation of those phenomena. Dewey thought it was possible, and absolutely crucial, to begin by describing what he called "generic traits of existence," because he thought such patterns were recurrent features of experience that indicate the conditions influencing everything we experience, think, say, and do.

Based on Dewey's nondualistic, process-oriented account of experience, we then have to explain how we can go beyond situations where things and events just *happen* to us and affect us, to the point where those things and events acquire meaning and can be taken up into processes of thinking. We therefore need an account of how it is possible to move from the plateau of inanimate material events to the plane of living organic processes, and eventually up to the emergent functions of mind, thought, and language. This move up to the level of meaning is a huge development, often tied to the emergence of language, so Dewey understands mind as a system of shared meanings, or possibilities for meaning, that emerge in a context of social communication.

Meaning is often regarded as exclusively linguistic, arising from communicative practices rooted in a specific natural language. Although Dewey emphasizes the importance of linguistic practices in constituting systems of shared meaning, he also saw that there is nonlinguistic meaning that is present before language acquisition and that reaches beyond our linguistic capacities and achievements. Embodied meaning is one of the hallmarks of a naturalized view of mind, thought, and language. Meaning rooted in biological and social processes becomes the basis for meaning as it operates in abstract conceptualization and reasoning (Feldman 2006; Lakoff & Johnson 1999).

The emergence of these capacities for making and experiencing meaning goes hand in hand with the emergence of mind, which Dewey regards as a system of socially and culturally shared meanings that make possible human communicative interactions. Mind is thus inherently embodied and social, and it is only derivatively individual (Johnson & Tucker 2021; Tomasello 1999). What we traditionally think of as individual minds (or individuals with minds) are actually just occasions of an organism's participation in historically developing cultural systems of meaning.

The acquisition of shared meaning enables thought, conceptualization, and reasoning. However, from a naturalistic perspective, these traditionally "cognitive" faculties are never purely formal or disembodied operations. Instead, they are deeply and extensively rooted in bodily (i.e., sensory, motor, and affective) structures and processes (Di Paolo et al. 2018; Swanson 2003a). Since mind and body are not two ontologically different kinds of reality, all our abstract thinking has to be grounded in bodily processes, operating with body-based meaning (Feldman 2006; Johnson & Tucker 2021). We recruit these body-based structures and processes for "higher-level" abstract conceptualization and reasoning (Lakoff & Johnson 1999; Schulkin 2000). Abstraction is not an act of transcending the body, but rather an appropriation or exaptation of bodily processes for new uses in new contexts (Gallagher 2005; Johnson & Tucker 2021, chapters 9–10). There is an ontological and epistemic continuity between our mostly nonconscious sensory and motor sources of meaning and our more conscious and reflective cognitive operations. Once again, all of our capacities grow from our embodiment, through brain, body, and environment (Schulkin 2004). Thinking is an action that transforms experience by means of new meanings entertained and used as tools for inquiry and growth of understanding. None of the so-called faculties of mind are either discrete faculties or purely formal operations (Dewey 1925/1981; Tucker 2007). They are not entities of any sort, but rather emergent functions for the remaking of experience, and they are *all* embodied.

With thought being understood as a transformative functional process, knowing becomes an activity of reworking experience through inquiry, rather than some mirroring of a pre-existing mind-independent state of affairs. Knowing is making a new experience, instead of reporting on a premade one. It is not an exclusively internal mental operation, but a form of action reaching out into nature and transforming it. Hence, it is always a process of learning the meaning of things and events. Cognition is learning extended over one's lifetime (Tucker 2007). Knowing occurs through the resolution of indeterminacy in a problematic situation, so as to open up more satisfactory functioning in the world (Berlyne 1954; Loewenstein 1996; Miller 1959). Knowledge—as a noun—is a misleading term, since there is never a finally fixed or finished state of knowing or body of knowledge. Knowing—as a verb—signifies the more or less successful reduction of indeterminacy in a problematic situation (Johnson & Tucker 2021). As such, any claim to knowledge must always be subject to critical reconsideration, in light of newly arising conditions.

If mind is a product of social communicative transactions, then how do we understand consciousness within a naturalistic framework? To state the obvious, it cannot be the intrusion of some transcendent mental operation (of a transcendent ego) into the world. Instead, it too must be an emergent function. Dewey treats consciousness as the ability to feel, with awareness, changes in one's body-state, in relation to the organism's ongoing interaction with its surroundings (Damasio 1999, 2010). Consciousness is not a "thing" or "entity" of any sort, but the feeling of qualitative characteristics of situations (Edelman & Tononi 2000; James 1890).

The profound role of qualities, feelings, and emotions in all cognition introduces one of the most distinctive and radical of Dewey's views: namely, that experience is massively shaped by what might be called our aesthetic sensibility. In response to the traditional marginalization, or even denigration, of aesthetic aspects of experience and cognition as being subjective and feeling based, Dewey brings aesthetics front and center in all experience. He thereby reminds us that the basic components and aspects of all meaning are fundamentally aesthetic. Feelings and emotions are not merely subjective private organic processes but rather the shared means by which any object or event can become meaningful to us. Living a meaningful and purposeful life thus becomes an aesthetic and artistic undertaking (Dewey 1934a/1987).

The previously sketched account of embodied experience emphasizes the pervasiveness of values in everything we experience, think, and do. From our opening account of experience, as rooted in organism–environment interactions, we saw that we are motivated at the deepest levels of our being—in basic biological viability and functioning—by organismic values. From a naturalistic perspective, it is values all the way down, and all the way up. All cognition is motivated. All thought is motivated. All knowing is motivated. The motive control mechanisms for basic bodily perception and action continue to operate at the highest level of our thinking and doing (Engel et al. 2015; Tucker & Luu 2012).

We are valuing creatures. Our values arise from our evolutionary development, coupled with our individual development over our lifetime. That we have certain values (biological, interpersonal, and cultural) is a demonstrable fact. However, recognizing which values (as organic habits) are currently driving our thinking and doing is only the beginning of intelligent living. We humans are marked by our ability to reflect on the meaning of our values, and we are able to criticize values, as a basis for transforming our value-based habits. Dewey regards metaphysics as supplying what he called a ground-map of our habits and values, and he regards philosophy as pursuit of the wisdom for living that comes from intelligent management, criticism, and reconstruction of our values. The result is a view of philosophy as the application of human reflective and critical capacities for the intelligent management of our lives. Dewey thus provides a scientifically and philosophically sophisticated naturalistic view of the human condition that provides a critical perspective for the ongoing assessment of our value commitments. In light of newly emerging conditions, we are able to reconstruct our embedded habits of thinking, valuing, and doing. His naturalism gives us an optimistic outlook, one rooted firmly in the science of human nature and one that is psychologically realistic. It tells us that all the values available to us, and all the possibilities for who we may become, arise from our embodied creaturely nature. Within such a framework, we can find meaning and purpose, not in eternal transcendent values, but in the use of our embodied intelligence for a life well lived.

2 It All Starts with Experience

The Importance of Beginnings

Beginnings matter. Every scientific project, every philosophical orientation, and every artistic endeavor has to start somewhere and somehow. In constructing a theory, the beginning makes all the difference, because where one begins determines what the relevant phenomena are, and how they are to be circumscribed and examined. For example, accounts of mind and knowing have traditionally picked as their starting point things like innate ideas, atomistic sense impressions, sense data, neural activation patterns, qualities, feelings, and so forth—take your pick. If, for instance, you begin with atomistic sense impressions, your main problem gets defined as how atomistic sense givens get combined, via association, into larger wholes and how they eventuate in higher cognitive operations. When you select your preferred candidate from such a list of phenomena, you are already employing values about what matters most and about what dimensions are relevant for explaining those phenomena. Moreover, in choosing where and how to begin, you establish a preferred method of inquiry and explanation. Different theories are distinguished by their different methods of inquiry, such as analysis of intuitions, behavioral analysis, phenomenological description, and neuro-imaging results.

To start, you have to select what you think the relevant phenomena are. Are they macroscopic physical objects, qualities of objects, subatomic particles, or even mathematical entities and relations? It is a value-based decision one has to make about *what* it is that needs to be explained, and then the question arises about what type of account is appropriate for the selected phenomena.

Dewey lamented that so much of traditional Western philosophy was the result of selecting certain "objects" (such as things, qualities, relations) as the focus of inquiry, and then pretending that those objects, qualities, or relations are all that matters for a comprehensive account of nature. The mistake is "the bias toward treating objects selected because of their value in some special context as the 'real,' in a superior and invidious sense" (Dewey 1925/1981, 33). The problem is not the need for selection. All inquiry is selective, and all selection is biased, that is, perspectival and based on preferred values. The problem arises only when we take what we have selected and regard it as the be-all and end-all of reality—as that alone which determines the nature of things. This selective emphasis only becomes a mistake when we pick out something that, for some reason, is useful and matters to us, and we then pretend that what we've selected for attention constitutes *the being* of things in their essential natures. Dewey speaks of the "transformation effected by philosophers of the traits they find good (simplicity, certainty, nobility, permanence, etc.) into fixed traits of real Being" (1925/1981, 33) The result is that we overlook the vast depth, breadth, and richness of human experience. The unhappy consequence of this reductive selectivity is, at best, a very partial grasp of the fullness of what is experienced and how it is experienced. At the worst, we ignore aspects that might eventually turn out to challenge our limited perspective and to open up a new mode of inquiry. We will return to this theme in subsequent chapters.

So, where does Dewey think we ought to start, if we have any hope of avoiding the bias and limitations of any particular method of inquiry or the fallacy of selective emphasis? Dewey's answer is: We must start with *experience*—experience in its fullest, richest, and most comprehensive sense. Our inquiry ought to arise from our mundane experience—where we live, move, and have our being.

In choosing the philosophically and psychologically loaded term *experience*, Dewey was painfully aware that this term carries with it a halo of problematic meanings and implications that must be identified and avoided. The single biggest problem is the mistaken assumption that experience marks out the subjective, internal *experiencing* of an individual person, which is then set off against the objects or events experienced as "out there" in the mind-independent world. As Dewey says, "When objects are isolated from the experience through which they are reached and in which they function, experience itself becomes reduced to the mere process of experiencing, and experiencing is therefore treated as if it were also complete in itself. We get the absurdity of an experiencing which experiences only itself, states and processes of consciousness, instead of the things of nature" (Dewey 1925/1981, 21).

The fundamental mistake is to think that our access to objects in the world is always and only through inner states of a subjective private consciousness. If all we know is our *experiencing*, then how could we ever know our world? Locked up in our subjective consciousness, we encounter the false problem of how something "inner" (e.g., an idea, perception, concept, or proposition) can ever relate to the "outer" (e.g., some objective state of affairs in nature).

Dewey's answer to this false internal versus external dilemma is to assert that we do not experience our experiencing, but rather we experience objects and events in the world:

> [Experience] includes *what* men do and suffer, *what* they strive for, love, believe and endure, and also *how* men act and are acted upon, the ways in which they do and suffer, desire and enjoy, see, believe, imagine—in short, processes of *experiencing*. "Experience" denotes the planted field, the sowed seeds, the reaped harvests, the changes of night and day, spring and autumn, wet and dry, heat and cold, that are observed, feared, longed for; it also denotes the one who plants and reaps, who works and rejoices, hopes, fears, plans, invokes magic or chemistry to aid him, who is downcast or triumphant. (Dewey 1925/1981, 18)

Experience is thus our way of being in, and being in touch with, our world. It enters into the ongoing course of events, by way of perception, thought, and action. Yes, experience includes our doings and sufferings, but it also includes *what* we experience and *how* we experience it. Though Dewey's most important book is entitled *Experience and Nature*, he soon realized how that title misleadingly suggests that experience and nature are two different things, which would then somehow need to be reunified. His rejection of any such dualism leads to his claim that, instead of being a veil that hides nature, experience is our only access to nature, as it reaches down into natural processes. Dewey urges us to reconceive the relation of experience and nature, so that "nature and experience get on harmoniously together—wherein experience presents itself as the method, and the only method, for getting at nature, penetrating its secrets, and wherein nature empirically disclosed (by the use of empirical method in natural science) deepens, enriches and directs the further development of experience" (Dewey 1925/1981, 10–11).

The way to avoid a noxious unscientific dualism, is to see experience and nature as dimensions of one continuous process of organism–environment interactions. We don't have nature plus an experience of nature. We simply have experience, which is both *of* and *within* nature: "Experience is *of* as well as *in* nature. It is not experience which is experienced, but nature—stones, plants, animals, diseases, health, temperature, electricity, and so on. Things interacting in certain ways *are* experience; they are what is experienced. Linked in certain other ways with another natural object—the human organism—they are *how* things are experienced as well. Experience thus reaches down into nature; it has depth. It also has breadth and to an indefinitely plastic extent. It stretches. That stretch constitutes inference" (Dewey 1925/1981, 12–13).

Dewey's alternative to the irreparable dualisms of mind versus body, mental versus physical, cognitive versus emotive, fact versus value, and thought versus feeling, is his radical naturalistic nondualism, which treats what were previously thought of as different kinds of reality as, instead, emergent functional organizations of the processes of an organism interacting with its environment. We are reminded just how easy it is to fall prey to dualistic thinking, if we notice that even Dewey's most famous starting point—organism–environment interaction—inclines us toward the dualism of two separable and functionally independent things, an organism and its environment. The antidote to this mistake is recognition that there is no organism without an environment, and, strictly speaking, no environment without an organism for whom it is an environment.

One might object by pointing out that there would still be a world of physical objects, processes, and events, even if there happened to be no organisms in that world. That is true, but it is equally true that in such a world, there would be no environments, since there would be no organisms for whom certain patterns of energy could constitute *environments*. That is why Jacob Von Uexkull (1957) coined the term *Umwelt* to indicate the meaningful affordances provided to an organism in certain surroundings. From this perspective, one could say that different animals, with different bodies and different sensory and motor capacities, might have different *Umwelten* even as they share the same environment (Lorenz 1970; Tinbergen 1951, 1963; Von Frisch 1953). We will elaborate on this point later, but the key for now is that organism and environment are two dimensions of one unified, continuous process that is experience.

The way to regard experience and nature nondualistically, then, is to see experience as our way of inhabiting—being at home in and with—nature. Experience is our access to natural processes, and it reveals their meaning.

Experience and Empirical Method

The supreme challenge for a naturalistic approach is to describe and explain, via empirical methods, the phenomena of nature (both nonhuman and human alike), in a way that does justice to the full richness and complexity of experience. No easy task! Dewey warned that we tend to select one significant aspect of experience that interests us and then we ignore everything else, thereby overlooking the inclusive richness of experience. Dewey argued that the antidote to this fallacy of selective emphasis, which is so prevalent that he called it the Philosopher's Fallacy, was to identify general structural characteristics of all experiences that establish the conditions for what we can experience and how it is experienced. He thus proposed what he called an *empirical metaphysics*, "as a statement of the generic traits of existences of all kinds without regard to their differentiation into physical and mental" (Dewey 1925/1981, 308). As we will see later, he gave examples of generic traits such as "qualitative individuality and constant relations, contingency and need, movement and arrest" (1925/1981, 308), and asserted that a suitable empirical method is the only way to appreciate the full range and importance of these general features of experienced things. Most of the errors of philosophy, leading to false philosophical problems, are the result of failure to include all of these recurring dimensions in one's metaphysics, which Dewey called a "ground-map" of existence (see chapter 9). An empirical metaphysics is proposed as the naturalized, empirical way of surveying recurrent general patterns of experience that are meaningful to us.

Richard Rorty (1982) challenged Dewey's project, providing two criticisms of the idea of an empirical metaphysics. The first is that an empirical metaphysics is an oxymoron, a contradiction in terms, because an alleged account of the general structures of all existences will be a foundational metaphysical account of just the sort Dewey went to great lengths to criticize. The second is that there is no distinctively philosophical method for getting at such traits, at least no method that wouldn't also be available to scientists, artists, and fiction writers. Let us consider each challenge in turn.

In the first argument, Rorty praises Dewey for his thorough dismantling of traditional philosophical dualisms and his demolition of any project that posits pure or absolute foundations for knowledge of any sort, but he then accuses Dewey of reverting to just such a foundationalism in attempting to craft an empirical metaphysics of generic traits. Rorty's argument, in this opening phase, boils down to the claim that no metaphysics is (or can be) a good metaphysics. The assumption is that any metaphysics is bound to be foundationalist, insofar as it will propose generic traits of *all* existences as the ultimate constituents of reality. So, says Rorty, Dewey is faced with the following dilemma: "Either Dewey's metaphysics differs from 'traditional metaphysics' in not having a directing bias concerning social values because Dewey has found an 'empirical' way of doing metaphysics which abstracts from any such biases and values, or else when Dewey falls into his vein of talking of the 'generic traits manifested by existences of all kinds' he is in slightly bad faith" (Rorty 1982, 74). Rorty correctly observes that Dewey cannot embrace the first horn of this dilemma, because that would amount to asserting that an empirical metaphysics alone somehow escapes "a directing bias concerning social values" (Rorty 1982, 74). Dewey's emphasis on the context or situation in which experience develops recognizes that any description of patterns of experience must rest on some set of values.

Rorty was an astute reader whose criticisms have to be taken seriously. If Dewey's "generic traits" were absolute category structures within nature, something like Aristotelian categories, or if they were a priori structures imposed on experience by the mind, like Kant's categories, then Dewey would be guilty of reverting to an untenable ontological and epistemic foundationalism. However, there is an alternative interpretation, according to which the generic traits are recurrent patterns manifested in existential events, but the meaning and force of these patterns depends on historical and contextual contingencies. Consequently, there is no context free or ahistorical structure of experience. For example, the first generic structural characteristic discussed by Dewey in *Experience and Nature* is that experience manifests both precariousness and relative stability. Experience is a mixture of indeterminacies and disruptions, but also of relatively stable patterns. However, the meaning, value, and significance of the precarious and stable dimensions will vary depending on the context. Therefore, Dewey avoids the foundationalism of fixed, pregiven, univocal characteristics of events, insofar as their meaning is context dependent. According to this perspective, Dewey is trying to get us to "look and see" what nature affords us by way of interaction and engagement. There are multiple experiential patterns that determine for us the meaning of what is happening at any particular point in time. From this perspective, the challenge for an empirical metaphysics is to be inclusive and pluralistic, never focusing exclusively on just one generic trait (e.g., stability) while overlooking another (e.g., precariousness). It is this situation-dependence that makes it possible for Dewey to avoid foundationalism and reductionism.

Dewey does himself no favors in this regard when he proceeds to make a major distinction between what he calls "primary" (macroscopic) versus "secondary" (reflective) experience. He introduces this distinction as follows: "This consideration of method may suitably begin with the contrast between gross, macroscopic, crude subject-matters in primary experience and the refined, derived objects of reflection. The distinction is one between what is experienced as a result of a minimum of incidental reflection and what is experienced in consequence of continued and regulated reflective inquiry. For derived and refined products are experienced only because of the intervention of systematic thinking" (Dewey 1925/1981, 15).

The important idea motivating this primary versus secondary distinction is that the process of critical reflection changes nature, rather than merely commenting on a pregiven, fixed nature. We may experience macroscopic objects via a mostly nonreflective sensory and motor engagement with our surroundings, but the introduction of reflective thought (say, via scientific or philosophical inquiry) provides a new grasp of the meaning of objects and events. Taken uncritically, it does seem that we encounter a world of macroscopic objects and events, and then scientific investigation can sometimes describe objects, forces, and causes that reveal nature in a different mode—that of causal sequences of events, some of which are not observable to the naked eye. Wilfred Sellars (1956/1997, 1962, 1968) would later call these two perspectives the *manifest* versus the *scientific* image of the world. For example, there's the chair as we see it, touch it, sit in it, and move it—a solid, hard, colored, functional object—and then there's the chair as we understand it theoretically via scientific experiment—a conglomeration of atomic and subatomic particles within mostly empty space. On this reading, Dewey is just trying to get us to appreciate that there are multiple meanings of "chair," relative to our values and practices of inquiry. The "chair" is what it is experienced as (Dewey 1905), and experience can include bodily engagement with a wooden, metal, or plastic object (e.g., sitting, lifting, moving), just as much as it can include scientific inquiry into what the chair is (e.g., its molecular structure) and how it enters into causal sequences.

As much as we might appreciate Dewey's motivations for the primary versus secondary dichotomy, it does not hold up well under scrutiny. What is the alleged distinction, really? Supposedly, the primary objects are the result of a "minimum of incidental reflection," but what does *that* mean? There may be some reflection, but just a "minimum?" Obviously, there is no hard and fast line here, but rather a sliding scale of increasing reflective intervention. Once an infant grows into childhood, the supposed objects of primary experience become repositories of sedimented meanings, many of which will be the result of reflective thinking. At the very best, then, the primary versus secondary distinction ought to be understood on a continuum, from the most minimal reflection involved in mundane experience of macroscopic objects up to objects and events regarded almost entirely through the lens of higher-level theoretical reflection. Objects of experience do not come to us with "primary" or "secondary" inscribed on them. In a given context one person's *primary* might be another person's *secondary*.

In spite of the impossibility of rigidly demarcating the primary from the secondary, we suggest that the distinction still does some useful work. It helps us realize that the meaning of things and events will depend on how we interactively engage them, whether at the level of bodily transactions with perceptible objects and events, or at the level of reflective thinking via various methods of scientific inquiry. Taken in this way, the distinction is trying to capture the need for a nonreductive, context-sensitive description of the richness of our lived experience. It is not meant as a fixed categorical framework. Dewey probably could have written *Experience and Nature* without this distinction, but then he would have had to find another way to make the distinction between objects engaged via sensory and motor interactions and those as understood within various scientific and philosophical frameworks. As we will see later, in the chapter on knowledge, the operations of inquiry, as secondary processes, remake experience by instituting new meanings. Consequently, there is a difference between an object prior to reflective engagement and that object as reconstituted via reflective inquiry. If the primary versus secondary scheme rejects any claim to being a priori, fixed, or absolute, then it becomes serviceable for helping us avoid the fallacy of selecting certain characteristics of a situation that we find useful or significant, and then treating them as fixed essences that capture everything that matters in experience. Instead, experience is plural and our understanding of it is multi-perspectival.

This brings us to Rorty's second criticism of Dewey's empirical metaphysics and his empirical denotative method. Rorty argues that there exists no "empirical method" that is the sole purview of philosophers, which is not also available to painters, poets, novelists, and musicians. Is there any need, Rorty asks, for some alleged discipline of thought that is the sole possession of philosophers? "Would anyone—including Dewey himself—really believe that there is a discipline that could somehow do for 'the basic types of involvement,' something left undone by novelists, sociologists, poets, and historians? . . . Nothing save the myth that there is something special called 'philosophy' that provides the paradigm of a synthetic discipline, and a figure called 'the philosopher' who is the paradigm of the intellectual, suggests that the professional philosopher's work is incomplete unless he has drawn up a list of the 'generic traits of all existence' or discovers 'the basic types of involvements.'" (Rorty 1982, 77).

Rorty makes an important valid point here. There is nothing unique about the role of the philosopher in constructing an empirical metaphysics that would set her apart from poets, novelists, and artists, when it comes to insight about experience. Dewey would emphatically agree. In the final chapter of *Experience and Nature*, he makes it clear that the philosopher is concerned more with discovering meaning than with discovering grand truths, in competition with scientific theories. Instead, the philosopher is a fellow traveler with anyone who has some way of exploring the meaning of things and events. As we will see in the last two chapters, what the philosopher brings to inquiry is not a unique method for generating truths, but rather experience in exploring the meaning of selected experiences, as well as training in analyzing underlying assumptions, appreciating the historical context of theories, and investigating the implications of certain distinctions for experience. Philosophers help us develop a synthetic overview of experience, but so do poets, novelists, and artists.

In summary, Dewey's "empirical method" is actually a modest phenomenology of lived experience, but minus any notion of purity and necessity of the sort championed by transcendental phenomenology. What is salvageable from the idea of an empirical metaphysics is the need to respect the complexity and richness of the situations we are describing and explaining. We need to find the recurring characteristics or patterns of experience that give us insight into our unreflective assumptions and prejudgments, so that we can, through empirical inquiry, explore the meaning of what we are attending to, without becoming reductive or hypostatizing. Thomas Alexander so nicely sums this up as follows: "The denotative method is meant to contextualize the *cognitive* interests of philosophy within the *noncognitive* scope of life. It is a method *for philosophy* that aims at *reminding* it of its *origins* in a pre-reflective world as well as of its *obligations* to make life more meaningful and value-rich" (Alexander 2013, 61; italics in original). However universal any such characteristic structures might be, what matters is the meanings of those patterns in concrete historical situations.

The Organism Striving for Viability in Its World

For humans, experience is an ongoing, eventful process of an evolving organism interacting with its changing environment. At the most basic biological level, experience is a transactional process, in which an organism appetitively exploring its surroundings seeks biological viability. In this way, the structures of our basic biological operations establish the basis for all our higher-level capacities and activities. A naturalistic approach must therefore always begin by describing the recurring structural characteristics arising from the organism's perception, bodily movements, feelings, and values. It then has to explain how these biological patterns and processes establish the conditions for higher-level cognitive activities, revealing a continuity between basic bodily functioning and abstract conceptualization and reasoning. We start, therefore, with the basic engagement of the animal with its surroundings, as the source of experience.

Within the framework we are developing, all things harken back to Darwin, who helped us recognize our inheritance, from our animal ancestors, of the basic body and brain architectures that underlie our functional capacities (Dobzhansky 1962; Gould 2002; Gould & Eldridge 1977). Recognizing the key role of ongoing evolutionary change, Dewey (1910) understood that philosophy was going to be different post-Darwin. The question was then, and still remains, how to integrate biological explanations into the larger culture in which we are situated. As animals, we are part of nature, even as we create diverse forms of culture. So, how does our animal evolution explain the possibility of our most impressive cultural achievements? This is a developing theme throughout this book, but we begin here with an account of the general patterns of our biological functioning.

Wallace Craig (1918), American psychologist, rooted the study of animal behavior in the natural world. This strain of research, later developed more fully by the Noble Prize winners Niko Tinbergen (1951, 1963) and Konrad Lorenz (1970), would become known as "ethology"—the study of animal behavior and social organization from an evolutionary biological perspective. Very much under the influence of Darwin, the investigation of animal behavior, physiology, and neuroscience would be understood in light of capabilities, evolutionary development, duration of life cycles, and individual development (Gregory 1966; Griffin 1974).

An animal was understood in the context of its environment—an environment where success was measured by viability and reproductive success. The first order of business is an organism's survival, and the second order is enhancement of the organism's well-being, or flourishing. The degree of complexity is a feature of the species, the niche, and the flexibility of the endogenous systems in engaging the changing environment. Animal development is seen in relation to problem solving, tool use, expanding capabilities, and conservation of energy.

Craig was interested in appetitive attraction/aversion and consummatory events. This was seen in the context of motivational allures, such as which needs are more pressing, and need attending to *now*, versus those which can be put off as less urgent. With respect to the motivational buildup of energy within an animal, needs that warrant satisfaction are higher on the appetitive search list, constituting goals to be attained and consummatory responses satisfied (Simon 1982).

Dewey (1925/1981) saw this need–search–satisfaction process as being the most basic pattern of organism–environment transaction, a continuous calling forth of active search in one's surroundings (Dewey 1896). When appetitive search is successful, and present needs are satisfied, all goes smoothly, but when needs go unmet, there is a functional disruption, which initiates a search for new means of satisfaction. In this way, a recurrent pattern of animal behavior arises, in which the animal falls in and out of harmony with its surroundings, as manifest in the need–search–satisfaction process (Dewey 1934a/1987).

The process just described is ongoing over the life of the animal. New ends-in-view are continually arising from appetitive search. An end realized, a need fulfilled, becomes a starting off place for further ends and goals. The distal goal of these search activities is animal viability—not some perfect state, but rather a "good enough" satisfaction relative to the animal's capabilities, needs, and energy reserves. Starting points and end points (which become further starting points) establish the context for the active probing, making, and doing that is guided by attractions and aversions arising from the search for resources, shelter, and safety.

The root metaphor operating here is one of activity—active doing and making—not passivity. An animal is a cluster of impulsions organized into habits of actively perceiving, manipulating objects, and moving in the world. Even perception begins with active exploration of one's environment (Dewey 1896). What and how we perceive something depends on experientially acquired expectancies projected out into the world, and then tested in relation to what actually transpires in experience at the sensory–motor interface with the animal's surroundings (Johnson & Tucker 2021).

The character of our active searching and striving reveals our most basic biological values. At this biological level the ultimate motivating value is *viability*—staying alive and at least minimally functional, given our capacities, needs, and available resources. The desired situation is satisfaction of needs with minimal expenditure of energy. What we strive for is shaped by finite resources externally (i.e., limited environmental affordances), and finite resources internally (i.e., energy reserves). These are recurrent themes across the life cycles, operative not just in biological self-maintenance, but, as we will see (chapter 6), operative equally in higher cognitive functioning, such as conceptualization and reasoning.

Approach, avoidance, and consummatory behaviors have long been noted in both vertebrates and invertebrates, such as social ants (Dethier & Stellar 1961; Maier & Schnierla 1935/1964) and flies (Stellar & Stellar 1985; Von Holst 1973). You don't even need a cortex to engage in this life-sustaining activity. Bees, for example, have no neocortex, and no older cortical features, and yet their range of capability is quite impressive. They are highly communicative with each other, and they orchestrate strikingly anticipatory behaviors in seeking resources such as nourishment and safety. Invertebrates, like bees, secrete diverse informational molecules (e.g., peptides, transmitters) in the organization of appetitive and consummatory experience, especially in anticipation of rewards (e.g., Keverne & Curley 2004, 2008; Taghert & Nitabach 2012). This design reaches back a half a billion years, and it underlies diverse forms of appetitive and consummatory behavior found in animals generally.

And then, evolutionarily, something new emerges, something which goes beyond mere satisfaction and the distribution of needed resources (e.g., glucose, protein, sodium, calcium) to bodily organs! With evolved complexity comes the anticipation of aversive or avoidance events, and the need to manage danger. Most encounters among animals are mixed. Elements of danger accompany achievement or satisfaction of one's needs and desires. A trip to the waterhole is not a free ride toward consummatory satisfaction (a nice drink of water). Danger lurks for the gazelle and the macaque, which must continually anticipate possible threats. Here, as has so often been observed, it is far better to respond to a false positive concerning threat, even if it expends valuable energy, than to succumb to a false negative, which results in being eaten. Hence, the neuropeptides dedicated to being alert to and combatting danger or pain (e.g., enkephalin, corticotropin-releasing hormone) are released, as are the resources dedicated to the conservation of water (e.g., vasopressin). Managing alertness and energy conservation are commonplace husbandry. We remember what available resources are situated in various places, in the context of attending to diverse needs. In social species like us, this also applies to considerations of who we are with and who we can rely on, predict, and utilize toward common ends, such as safety, alertness, comfort, recognition of alliances. All of this plays out within contexts of search for satisfying consummatory experiences (Berridge 2004; Schulkin 2017; W. J. Smith 1997; Stellar & Stellar 1985).

Humans, and some other mammals, can sample from their environment and learn about the likelihood of certain events. This is not limited merely to searching for resources to satisfy present needs but extends to the capacity to anticipate the likelihood of future experiences. Certain animals can recognize whether certain events are linked to appetitive ends and provide consummatory satisfaction, such as recognizing where salt can be found, even when we are not currently in need of it. They then utilize that information when needed in the future.

In this appetitive search process, one important adaptation has been to keep memory storage external, and with easy access (Clark 1998; Clark & Chalmers 1998). So, information storage is not strictly in the head, but off-loaded into the environment we are living in and adapting to (Donald 1991, 2004). Another core strategy is the metabolic monitoring of energy expenditure, in order to keep it as minimal as possible, or, at least, avoiding depletion of resources.

Homeostasis and Allostasis in the Maintenance of Life

Antonio Damasio observes that all life goes on within a boundary: "Life is carried out inside a boundary that defines a body. Life and the life urge exist inside a boundary, the selectively permeable wall that separates the internal environment from the external environment. The idea of organism revolves around the existence of that boundary. . . . If there is no boundary, there is no body, and if there is no body, there is no organism. Life needs a boundary. I believe that minds and consciousness, when they eventually appeared in evolution, were first and foremost about life and the life urges within a boundary" (Damasio 1999, 137). The boundary must be permeable, so that the organism can take in energy resources and expel waste. Consequently, every organism must have evolved structures dedicated to the maintenance of the conditions of life within the boundary of the physical organism. Viable organisms must be able to monitor their internal states and regulate the conditions necessary for life. The two most fundamental processes for achieving this state are homeostasis and allostasis. Homeostasis is about staying relatively the same, maintaining a set point of one kind or other. There are definite parameters for the range of oxygen, iron, glucose, calcium, and so forth, that can be tolerated. If your blood oxygen saturation drops below a certain level, you will die. If your salt level falls below a certain threshold, your animal functions will be diminished or even eliminated. Within these general parameters, however, there is considerable variation among individuals, reflecting their specific context of need, availability of resources, and level of threat or emergency.

The tradition that began in the middle of the nineteenth century, mostly associated with Claude Bernard (1865/1957), emphasized the regulation of the "internal milieu." Bernard sought to render physiology more like physics, the physics of Newton, which emphasized the necessary, law-like features of systems. There were features of deterministic set points. One focus for Bernard was glucose metabolism, in the context of diabetes. Change in glucose levels might require an urgent return to a prior state of internal stability. As usual, context matters, so the "correct" pre-established range of suitable values might vary from individual to individual, depending on specific conditions. The stability of a developing animal, or a human child, is not the same as an adult, a lactating female, or older, less active individuals. Life cycles, other needs, and need for greater conservation and distribution of energy are critical factors in the regulation of the internal milieu.

Bernard, influenced by a certain Newtonian conception of science, was focused on the deterministic features of physiological regulation. Consequently, in his theory, strict regulation dominates, and it was strict regulation mostly independent of regulatory contexts. He, therefore, downplayed contextual considerations where adjudication of needs, importance, and severity are computed and integrated in the organization of appetitive and consummatory experiences.

However, strict regulation is a luxury of the laboratory. It lacks the real-life complexity of regulatory adjudication in adjusting to changing circumstances and needs in the fields of life. Walter Cannon (1915/1929, 1932/1963) took the regulation of the internal milieu further by emphasizing "homeostasis" across a wide array of biological systems. Although Cannon did not coin the phrase "wisdom of the body," much of his work emphasized the diverse physiological systems that are essential for maintaining homeostasis, as the basis for the overall well-being of the organism. Consider, for example, set-point variables like glucose. Deviations in ingestion or conservation are met with a whole-body defensive response. In the case of glucose, it was known that insulin is essential for stable glucose regulation and metabolism. Hormones tied to absorption of nutrients and their redistribution were critically involved in the regulation of the internal milieu. Investigators discovered a great range of physiological mechanisms for sustaining blood volume, calcium, iron levels, and so on. The focus was on providing stability to the internal milieu, which is essential for continued viability.

The problem with this classic account of homeostasis is that it is limited to a conception of body–environment disruption, as passively experienced, giving rise to activation of bodily mechanisms for returning to the set-point value. However, our bodies and brains are far more active and plastic than that! We are a probing, inquiring species, anticipating possible disruptions and prospectively preparing to adjust to changing conditions. We are active organisms, projecting expectancies for experience and making readjustments to our experiential predictions in the face of change (Johnson & Tucker 2021). We are not just reactive, but actively anticipatory. In this sense, we are intelligent problem solvers using tools that aid regulation in response to threats from possible predators, blood loss, depletion of energy resources, and so forth.

In response to these changing needs, diverse bodily mechanisms are activated in correcting the imbalance within the animal. This is the result of the increased complexity of capabilities and cortical evolution, making it possible for us to actively anticipate, and adapt to, new circumstances.

Cannon (1932/1963) realized that the regulation of the internal milieu was not merely a matter of internal physiological control systems, but also required behavioral interventions in order to enable the necessary adjustment to changing conditions. However, he still saw this as a matter of maintaining a pre-established set point, in a decontextualized way that failed to appreciate ecological contexts that matter. Compromise and adjustment within limits are critical to continued viability and continued success in the animal's natural surroundings (Starling 1923). Cannon was still trapped within the confines of the idealized conditions of laboratory research, as opposed to the variability and trade-offs commonplace in natural settings.

What appeared next were important scientific discoveries concerning the behavioral regulation of the internal milieu. Curt Richter (1943) saw how bodily viability requires maintenance of specific chemical constituents (e.g., sodium, calcium, vitamin B) that are essential for bodily function and health. The hormones affecting fluid balance (e.g., angiotensin, vasopressin, aldosterone) act in the body to conserve sodium ions and water, and act in the brain to orchestrate the appropriate behavioral responses necessary for basic bodily functioning (Fitzsimons 1999). For example, in response to a bodily wound, the brain has various mechanisms for managing pain, retention of extra-cellular fluid, minimizing blood loss, and so on. Animals must therefore engage in active, exploratory behaviors geared toward acquiring those necessary nutrients and minerals. Richter thus recognized that homeostasis was not just an internal adjustment, but also depended on bodily appetitive and search behavior in the external world.

It is this sense of active regulation of the internal milieu and the social milieu that would begin to stretch the concept of homeostasis, taking us beyond just a reaction that initiates a return to some pre-established value. Richter (1965/1979) and others studied the circadian (twenty-four-hour) clock in the timing of hormone releases and behavioral responses. They explored the many clocks (twenty-four-hour, weekly, monthly, seasonal) that generate anticipatory physiological and behavioral responses toward what might be needed to remain viable, such as seeking shelter, warmth, or social contact. The mechanisms of the clocks reveal how the brain is active and logged into external events, such as light, darkness, seasons (Nelson et al. 2002; Wingfield 2004). Knowledge of this sort is essential in the organization of appetitive and consummatory events. The development of expectations directed toward external worlds, actively anticipating forthcoming events, constitutes a running theme of neural/behavioral regulation.

The point is not to be passive, but to limit the energy costs of a brain actively engaging an environment through bodily interventions. This research revealed the active, probing, inquiring brain and body projecting expectancies and regulating the body's response to its changing environment. Some animals get ready for seasonally cold weather by building nests and storing food in advance of future needs (Tinbergen 1951). These are common events, across diverse species, who need to plan ahead for future contingencies. For example, among elephants, pregnant or dominate females often direct younger elephants into caves where they come into contact with mineral deposits essential for lactation, bone growth, and systemic health (Denton 1982, 2005; Redmond 1982; Schulkin 1991).

In animal problem solving, geared toward life maintenance and well-being, the solution need not be ideal or perfect, but only "good enough," that is, achieving an appropriate balance of required nutrients, energy expenditure, and degree of satisfactory restoration. Perfection is mathematical, not biological. The issue about regulation is to link the active modes to life concerns. This involves getting ready for what might be needed, not just reacting to what is currently needed.

Ivan Pavlov (1897/1902, 1927/1960), known mostly for his behaviorist views on learning, actually understood learning as far more projective and anticipatory than mere stimulus–response behavior. Instead of an inert, passive, receptive body, Pavlov saw the body as always actively probing its environment. This search process seeks to anticipate what will be needed for continued viability. This will involve, for example, secretion of hormones, like insulin, that is required for maintaining requisite glucose levels, and for redistributing glucose, in the right amounts, to specific areas of the body (Power & Schulkin 2008; Powley 1977, 2000).

The term *allostasis* was coined to capture this anticipatory, plastic, and projective activity within and of the body. Homeostasis remains an important value for animals, but insofar as it focuses on returning a system or value to a preset level, it does not encompass the full range of preparatory and anticipatory bodily processes that allow animals to deal with changed and changing conditions.

At the neural level, allostatic regulation emphasizes not just conservation of neural resources (Sterling 2020), but neural genesis, the growth of new neural tissue essential for viability (McEwen 1998; 2007). Regulation dominates the neural space, anchored to changing contexts. Regulation is not just reactive and set-point driven, but recognizes varying set points, depending upon the context and the animal's adaptive capabilities. This pertains to both the basic bodily needs and also to social needs (e.g., closeness to others, cooperative behavior, shared goals). This anticipatory regulation depends on assessment of events, temporal cycles, spatial location, and probability of events. This is an ongoing active undertaking, turning on some systems, turning off others, and idling others—all in response to changing conditions. When regulation breaks down, the result is allostatic overload, pathology, and devolution of function (McEwen & Stellar 1993). One price of evolved capabilities that are emboldened and embodied in expanding contexts is exhaustion and eventually neural/behavioral breakdown. For example, one typical feature of aging is undue duress without regulatory relief, leading to allostatic failure.

Allostasis is a major evolutionary advantage for certain animals, and especially for humans (Sterling 2004; Sterling & Eyer 1988). It takes us beyond automated response to passively received stimuli. With allostatic capacities, we acquire a flexibility in reaching out to the affordances of our environment, as a way to manage change in body states, in response to changing environmental conditions (Schulkin 2003, 2011). We acquire the gift of anticipatory regulation of our internal milieu and, eventually, our social milieu. Antonio Damasio has recently suggested the single term "homeodynamics" as covering both homeostasis and allostasis: "Homeodynamic systems, as is certainly the case with living systems, self-organize the operations when they lose stability. At those bifurcation points, they exhibit complex behaviors with emergent characteristics such as bistable switches, thresholds, waves, gradients, and dynamic molecular rearrangements" (Damasio 2018, 52).

The Experiential Basis of Life Functions

The upshot of this chapter is that the starting point of all experience is our embodied animal interactions with our various environments. The chief value governing these interactions is viability in a world of danger and limited resources—a world at once material and social. We engage in appetitive search for consummatory experience, or satisfaction, consistent with our need to preserve energy resources, where possible. We share with many other animals some of the capacities and bodily architectures that shape our search activities related to survival and life maintenance. Two of our most important capacities (or values) are homeostatic restoration of a prior state, and the dynamic, anticipatory, predictive activity of allostasis. The latter allows us to achieve a flexibility of response and planning that goes beyond the preservation of a homeostatic state.

From a naturalistic perspective, our explanation of all human experience—perceiving, feeling, conceptualizing, reasoning, valuing, planning, and acting—has to be rooted in these basic bodily processes for survival and the enhancement of our well-being. One key to these achievements is recognition that we are active, searching, striving creatures, moved mostly by nonconscious processes and motivations, but also capable of more reflective evaluation, planning, and controlled action.

A naturalistic account cannot draw on supernatural or transcendent objects, forces, or values. It has to explain human behavior wholly in terms of emergent functional capacities and activities that arise through increasing levels of complexity in the organism (here, a human body) and in its environments (physical, interpersonal, and cultural). It is to that emergentist metaphysics that we now turn.

3 The Naturalized Metaphysics of Emerging Mind

The Need for a Naturalistic Emergent Metaphysics

Most people think of themselves as having a body and a mind. The body supposedly provides our sensory perception of our world, and it is the means of our causal action in that world. Our mind provides our awareness of things and their meaning for our lives, it enables us to plan and execute actions in the world, and it is the locus of our reasoning. Thus, there arises that classical philosophical question concerning whether mind and body are two kinds of realities, just one, or none at all.

Even for those of us who reject the strong dualism of body and mind as radically different types of being, there nevertheless sure seems to be something special about our capacity for conscious awareness and thought. For each of us, there is something it feels like *to be us*, here and now in this present moment. For the academic in her office, it might be the crisp sunlight of a dying winter's day, radiating in through the window to illuminate the surroundings and instruments of her daily life—her desk, chairs, bookshelves, books, computer, and favorite painting. For a tree faller, it might be trees, rocks, sky, mud, the roar of chainsaws, choker chains, fallen logs, hard hats, the smell of earth and forest, and the scent of cigarette smoke and chain bar oil.

Those who believe they have an eternal soul, or at least a disembodied ego that is only temporarily lodged in a finite body, have a ready-made seat for consciousness, concepts, reason, and will. Logic is taken to consist of the universal patterns of a transcendent reason. Knowledge is the ability to grasp dispassionately, through concepts and propositions, the objects, properties, and relations that make up the mind-independent world. Moral deliberation is the bringing of particular situations under rational, universally valid moral principles. And last but not least, a transcendent ego holds out the possibility that the death of your earthly body is not the last word, does not cast the final stone.

However, if you think that you are fundamentally just a highly complex social animal living out its fragile, finite existence, then you've got some explaining to do. You have to explain how our evolutionary history has supplied us with neural and bodily structures responsible for various sensory, motor, and cognitive functions, and then you have to explain how patterns of neural connection are strengthened or inhibited via your individual experience over the course of your life. Calling down transcendent entities, causes, or values is not an option.

Our project in this book is a version of this second, naturalistic option. We humans are animals—pretty remarkable animals—that are the result of the current state of our species' evolutionary history, plus the processes of our individual development over our life span. The most significant types of explanation available for these phenomena are the result of empirical inquiry and testing, supported by scientific research, phenomenological description, literary and artistic insights, and philosophical reflections on the human condition.

At first glance, dualistic theories may appear to have two distinct advantages over nondualistic naturalistic approaches. First, they fit right in with our commonsense understanding of ourselves as having a mind that controls our body, which fits right in with the idea of a transcendent ego that reasons, initiates causal effects in the world, and generates universal values. Second, hypotheses about transcendent egos can't be falsified, because they are not based on empirical inquiry and testing. For example, Kant argued that the only way to account for his intuitions about what he believed to be the universality, necessity, and unconditional bindingness of certain moral imperatives, is to postulate what he called a "pure practical reason" as the source of those absolute principles (Kant 1785/1981). While he admitted that he couldn't prove either that there is such a transcendent rational will (defined as practical reason), or that such a will is radically free, he argued that we need to postulate such a will, in order to explain the possibility of universal moral laws. This led Nietzsche (1878/1984, 1895/1990) to argue that Kantian rational morality is the last refuge for those who have given up God as the source of moral law and scrambled to replace Divine Reason with Universal Human Reason, as a way to shore up the crumbling edifice of religious ethics, with its supernatural foundations.

However, naturalistic theories are not without their own significant explanatory resources that come from a broad range of scientific disciplines. The fundamental challenge for naturalistic approaches is to explain the existence and operation of our many human capacities (e.g., life, perception, feeling, consciousness, thought, language, meaning, reason, knowing, aesthetic experience, and morality), without calling down supernatural or transcendent entities, agencies, forces, causes, or values.

From a naturalistic perspective, the key to understanding these sorts of phenomena is identification of emergent functional structures and processes that result from increased complexity of organisms and their environments. This amounts to the claim that, as organisms and their environments become more complex, through ongoing interaction and development, the environment comes to afford more possibilities for action and engagement. New functional organizations arise in the brain and body via transaction with the environment, first, to sustain life and enhance well-being, and, second, to make possible the shared meaning, values, and knowledge that underlie our cultural institutions and practices.

The supreme challenge, over and over again, is not just to *describe* emergent functions, but to *explain* how they are possible. For example, Dewey is an exemplary phenomenologist, by which we mean that he provides extremely nuanced, insightful, and rich descriptions of various functional capacities and diverse lived experiences. One of these, which we will examine shortly, is the difference between animals that are able to feel or discriminate something nonconsciously and other animals who possess the additional capacity to be consciously aware of their feeling response to some object or event. So, an animal that may be lacking in consciousness, might nevertheless be able to sense (and respond to) a change in temperature, light intensity, or glucose gradients, even though it doesn't *consciously* feel that discrimination or differentiation. This is feeling as sensitive discrimination of differences within an environment, coupled with the ability to respond to those. A different animal species might have evolved to the point where, in addition to this nonconscious mode of sensitiveness, they can consciously feel the change in intensity or quality. Even the most nuanced illuminating description of this emergent capacity of conscious feeling does not guarantee an adequate explanation of these phenomena so described.

Moreover, let's say that you go beyond mere description by proposing an account of the neural basis of a certain capacity or experience. Critics then object that neural explanation of a lived experience can never explain why the experience "feels" the way it does. Indeed, this problem about qualities of feelings has given rise to a minor cottage industry of articles and books on what has come to be known as the "qualia problem." A key issue in this literature is whether *any* scientific account could ever, under any conditions, adequately explain the feeling of the qualitative phenomena so described (Chalmers 1995; Dennett 1991).

We proceed, therefore, highly cognizant of these challenges to naturalistic explanations. In constructing an emergentist metaphysics, we strive to draw from a neuroscience that was not available in Dewey's day, but which now gives us new neuroscience tools for better explanations of some of the phenomena Dewey so well described. In some cases, we will suggest that a contemporary biological and neuroscientific explanation of emergent functions is all we can provide, and all that we need, in order to understand the relevant phenomena.

What we are proposing in this chapter is an emergentist process metaphysics, supported by contemporary biology and neuroscience, and therefore perhaps richer and more explanatorily satisfactory than Dewey's remarkable account, although following the general argument as Dewey developed it. Dewey identified three main organizational plateaus: inanimate matter (the physical), living beings (the psychophysical), and mind. We will generally preserve this rough tripartite functional organization, but attempt to give a more nuanced and differentiated account of how each emergent process operates, and under what conditions. The fortunes of an emergentist metaphysics are fateful, insofar as they affect the very possibility of a naturalistic perspective on the world.

From Inanimate Matter (Plateau 1) to Life (Plateau 2)

Philosophers interested primarily in higher cognition tend to focus on the nature of mind, concepts, thought, reasoning, meaning, and knowledge. Unsurprisingly, they often start by taking for granted the existence of life, and then explore how certain animals could be capable of consciousness, mind, thought, and language. Dewey starts, instead, with the development of life out of inanimate nature, not so much by explaining how that is possible, but mostly by describing what distinguishes inanimate from living things.

Dewey's basic claim is that, although inanimate physical processes and animate bodies are both part of webs of causal conditions, there is a major difference between the two. In the plateau of inanimate causation, there is no tendency of a thing to prefer one possible state over another, whereas animate creatures exhibit a tendency or need to preserve their existence and to maintain or restore a particular internal homeostatic or allostatic state: "Empirically speaking, the most obvious difference between living and non-living things is that the activities of the former are characterized by needs, by efforts which are active demands to satisfy needs, and by satisfactions" (Dewey 1925/1981, 194). In chapter 2 we described this basic need–search (effort)–satisfaction pattern from a biological perspective, as necessary for life maintenance. Dewey explains this organic pattern as follows: "By need is meant a condition of tensional distribution of energies such that the body is in a condition of uneasy or unstable equilibrium. By demand or effort is meant the fact that this state is manifested in movements which modify environing bodies in ways which react upon the body, so that its characteristic pattern of active equilibrium is restored. By satisfaction is meant this recovery of equilibrium pattern, consequent upon the changes of environment due to interactions with the active demands of the organism" (Dewey 1925/1981, 194).

This is Dewey's version of what we earlier described as the body's regulation of homeostatic and allostatic responses, in order to maintain or re-establish the conditions for the organism's survival and growth. Iron molecules, under the right conditions, such as the presence of oxygen, will become iron oxide (rust), but pieces of iron don't lie around wanting to grow up to become rusted iron. Iron has no end, aim, or goal (*telos*) that drives it toward a certain state. If the appropriate conditions are present, iron will do just what it does (what its chemical properties allow and give rise to, under certain environmental conditions). So, "in inanimate bodies as such, 'saturation' occurs indifferently, not in such a way as to tend to maintain a temporal pattern of activity" (Dewey 1925/1981, 195).

In contrast, living things seek to maintain at least minimal conditions for life. Plants, and even some unicellar organisms, do not have a central nervous system, but do have diverse chemical signaling systems (e.g., estrogen, insulin, endorphins) that regulate viability within the very limited confines of their modest capabilities and the kinds of environments they inhabit. Life is about regulation of organic processes, and the movement beyond the very limited capabilities of plants to the evolutionary emergence of simple animals was a game changer in terms of more complex systems for sustaining life.

Animals evolved brains and bodies that make it possible for the animal to monitor changes in its bodily state and to respond differentially to those changes, in order to generate or return to a certain dynamic equilibrium (i.e., homeostasis or allostasis). So, "with animate bodies, recovery or restoration of the equilibrium pattern applies to the complex integrated course or history" (Dewey 1925/1981, 195).

This need–search–satisfaction pattern is present even in the simplest of animals. Noted zoologist Vince Dethier (1968) demonstrated some years ago the length and degree to which blowflies that are hungry are subject to internal control systems, including olfactory and other sensory systems geared toward finding sources of nutrients. Olfactory, gustatory, and visual systems converge in guiding the fly to sources of nutrients. Volumetric and blood sugar monitoring systems regulate the restoration resulting in the satisfaction of the appetitive search. All this, even though "the brain of a blowfly weighs only about 0.84 milligrams (wet weight)!" (Dethier 1968).

This fundamental, life-preserving metabolic process works against entropy, the tendency of order to dissolve into disorder. Antonio Damasio has described this metabolic regulatory process with the Latin term *conatus*, "the continuous attempt at achieving a state of positively regulated life. . . . A blend of striving, endeavor, and tendency of each being to preserve itself" (Damasio 2018, 36). This is an appropriate description of a certain life pattern, as long as we avoid the error of turning this fact about how living things have evolved structures that enable life maintenance into a hypostatization of something like a "life force" (e.g., Bergson's (1946) *elan vitale*). There is no evidence for a mysterious life force that descends on some things (living things) and not others (inanimate things). Instead, certain types of complex bodies, which have, through evolution, found ways to monitor their body states and to make adjustments that preserve the integrity and growth of the organism, continue to exist—to live. When those conditions cease, the organism dies, but death is not a matter of a life force somehow exiting the body, leaving it lifeless. Instead, it is a matter of total functional breakdown.

In their book *Autopoiesis and Cognition* (1972), Chilean biologists Humberto Maturana and Francisco Varela coined the term *autopoiesis*, as a way of distinguishing living from nonliving (mechanical) systems. Ezequiel Di Paolo briefly summarizes autopoiesis as "the idea that a living system is one that is constantly constructing itself and by this activity making itself distinct from its environment" (Di Paolo 2018, 80). Recall, from chapter 2, Damasio's claim that life goes on "within a boundary," but a boundary that is permeable, since energy must flow into the organism and waste must be expelled from the organism. Di Paolo thus notes a basic tension in the original notion of autopoiesis, a tension between the need to establish a boundary that defines the living organism, and, at the same time, to allow energy flow across the boundary: "It is problematic to say that matter and energy may flow freely across the boundary of the organism, because if this happened it would soon become a violation of the self-distinction requirement due to uncontrolled transformative effects. On the contrary, pre-formed active matter and energy can only flow *conditionally* across the organismic boundary" (Di Paolo 2018, 84).

Thus, the conditions for life are tricky, and fragile. There is a delicate balancing act between maintaining a relatively closed boundary within which life goes on, and yet opening the boundary (keeping it sufficiently porous), as necessary, for transfer of matter and energy. The key is to continually monitor the state of the body, and its relation to its environment, in a way that maintains homeostatic and allostatic equilibrium.

From Sensing to Feeling

As we move from the plane of physical causal interactions to the plane of living organisms, there are several emergent functions that must be explained. As Dewey says, when we speak of rising to the psychophysical plateau, "the prefix 'psycho' denotes that physical activity has acquired addition properties, those of ability to procure a peculiar kind of interactive support of needs from surrounding media" (Dewey 1925/1981, 196). From a naturalistic perspective the idea that physical activity acquires additional properties suggests that we might look for these emergent properties in the neural, bodily, and environmental basis of continuity of functions that humans share with other animal species, and also in the continuities between so-called lower and higher functional capabilities within an organism.

The chief problem is how novel functional organizations become possible. Internal organization within an animal is the result of its evolutionary history and its adaptive response to conditions of its surroundings, over its life span. Organism organization results from the evolution of biological systems that support life maintenance and viability (Bernard 1865/1957). In the context of living things, specific sorts of functional organization make possible the continued existence of the organism. Walter Cannon argued that "the constant conditions which are maintained in the body might be termed equilibria" (Cannon 1932, 34), and this preservation of equilibrium is a brain–body–world coordination, requiring vital resources for sustained viability within different environments.

For example, in order to survive, grow, and flourish, animals must learn to respond differentially to aspects of their surroundings. This kind of discriminatory behavior is carried on, for the most part, without conscious awareness and feeling. Dewey concludes: "This pervasive operative presence of the whole in the part and of the part in the whole constitutes susceptibility—the capacity of feeling—whether or no this potentiality be actualized in plant-life. Responses are not merely selective, but are discriminatory, in behalf of some results rather than others. This discrimination is the essence of sensitivity" (Dewey 1925/1981, 197). Sensitive discrimination, both nonconscious and conscious, serves the organism's survival and flourishing. It is not always clear whether Dewey thinks sensitivity and feeling are two different phenomena, or aspects of one phenomenon. Sometimes he seems to use the term "feeling" to include both the nonconscious discrimination of features of objects or situations (which is what sensitivity is), and also the conscious feeling of qualitative differences. However, most of the time he uses "feeling" for the latter, that is, for awareness of the discriminated features or qualities: "'Feeling' is in general a name for the newly actualized quality acquired by events previously occurring upon a physical level, when these events come into more extensive and delicate relationships of interaction. More specifically, it is a name for the coming to existence of those ultimate differences in affairs which mark them off from one another and give them discreteness" (Dewey 1925/1981, 204).

The basic point here is that animals of our complexity can both *discriminate* features of our environment (i.e., we are capable of sensitiveness) and also *feel* some of these discriminations as qualitative differences. For example, consider certain one-celled animals that can detect in their ambient circumstances the glucose they need for metabolic processes, and are able to move toward higher glucose gradients in their surroundings. Here there is clearly sensitiveness, discrimination, and bodily response, but not conscious feeling. Animals of this sort don't *feel hungry*, but they very much need glucose and have evolved to search it out. They participate in need–search–satisfaction, though they don't know they are doing this. By contrast, humans, and certain other animals, sometimes consciously feel their needs and their satisfactions. They feel hungry, feel the need for nourishment, and feel satisfaction resulting from eating (W. J. Smith 1977; Stellar & Stellar 1985).

The term "feeling" is typically used for conscious awareness of both our qualitative discriminations and our emotions, but it is important to distinguish between feeling the flow of our sentient existence and feeling an emotional response (Langer 1937, 1953, 1972). We often have emotional responses of which we are not aware (as in, Jack: "Why are you so angry today?" Jill: "Who me? I'm not angry!" Jack: "Then why is your face red, why are your fists clenched, why is your breathing accelerated, and why is your jaw clenched?"). Emotional responses are mostly automated and unconscious behavioral patterns, but sometimes we are able to consciously *feel* those changes in our body-state in response to environmental cues (Damasio 1999). We feel certain moods and emotional states. We thus have to distinguish between what Damasio calls exhibiting an emotional response pattern and feeling that emotion: "Emotions are complex, largely automated programs of *actions* concocted by evolution. . . . Feelings of emotions, on the other hand, are composite *perceptions* of what happens in our body and mind when we are emoting. As far as the body is concerned, feelings are images of actions rather than actions themselves: the world of feelings is one of perceptions executed in brain maps" (Damasio 2010, 109).

Brains create *neural maps*, which are patterns of neural connectivity and activation generated by the body's perceiving and responding to aspects of its environment, both the external environment of objects and their configurations, and also our internal bodily states, accessed through proprioception, kinesthesia, and monitoring of visceral bodily processes. With respect to our external environment, for instance, there are retinotopic maps, in which relations of objects and qualitative differences in a visual scene are preserved in the patterns of neural connection in visual cortices. We also have tonotopic maps in which adjacent pitches are realized in adjacent neural ensembles or functional clusters. "The images in our minds are the brain's momentary maps of everything and of anything, inside our body and around it, concrete as well as abstract, actual or previously recorded in memory" (Damasio 2010, 70).

Emotions, in part, are bodily responses to modifications of our bodily (including brain) states arising from our interactions with our environments (Schulkin 2004). Emotions evolved to help us make important homeostatic and allostatic viability adjustments to our changing environment, as it is causing changes in our body-state. For the most part (though not always), emotional responses serve the well-being of the organism, helping us avoid dangers while pursuing satisfactions. As we will see later, emotions are therefore one of our most profound ways of discerning the meaning of our situation, even prior to, or without the use of, language and signs (Darwin 1871/1874, 1872/1998).

Feelings, then, "are experiences of certain aspects of the state of life within an organism. Those experiences are not mere decoration. They accomplish something extraordinary: a moment-to-moment report on the state of life in the interior of an organism" (Damasio 2018, 104). These ongoing assessments of the quality of life in the organism have valences, experienced as pleasant versus unpleasant, and all of the subtle variations in between those two poles of the continuum of emotional values. These feelings can range from pervasive moods (e.g., malaise to well-being) to the canonical emotions of fear, anger, surprise, joy, and so on (Cheney & Seyfarth 1990, 2007; LeDoux 1996).

The explanation of the conditions which make possible emotions and feelings will include both patterns of activated neural connections and also the release of chemicals that affect the state of the muscles and other bodily tissues. We can thus explain the basic conditions, the underlying values, and the purposes served by feelings and emotions. That is the limit of our present explanatory resources. Why prolonged contact of ice on your arm eventually leads to a burning feeling is quite explicable in terms of neural and bodily structures and processes, but why the burning feels just the way it does is not. However, we suggest that no one should expect more of an explanation than the sort just described. To observe that under bodily (including brain) and environmental conditions A, B, C, and so forth, you will feel a certain way (described phenomenologically in terms such as painful, pleasant, burning, stinging, cold, hot, warm, sleepy, irritable, joyful, and depressed), is all the explanation we can have, and all we need.

We will have much more to say about consciousness, feeling, and emotions later, but for now our goal is to indicate how to think about emergent functional processes in general. We must identify the specific conditions that make those functions possible and explain what values of the organism are served by the development of each particular capacity. We have to repeat this type of explanation all the way up to the highest manifestations of abstract conceptualization and reasoning.

From Life (Plateau 2) to Mind (Plateau 3)

The plateau of life culminates in living animals capable of self-organizing autopoietic activities that maintain the integrity of the bodily boundary, in a way that both separates the animal from its surroundings, while simultaneously keeping it actively in touch with that environment. More complex animals have neural maps for aspects of their external and internal environment (Damasio 2010, 2018; Gallistel 1982, 1993; Von Holst 1973). Via these maps, they discriminate features of these environments, which amounts to sensitivity. They have evolved systems for responding to these perceptual situations, both by changes of body state and actions in the world. These interacting systems are directed toward the viability of the animal in its surroundings. Those animals capable of consciousness can actually feel changes in their body state, arising from the animal's monitoring of its ongoing engagement with its environment.

However, having a feeling does not necessarily correlate with understanding or knowing what you are feeling. Dewey contrasted being in a feeling state that is "dumbly rapturous"—where you feel something, but aren't sure what you are feeling—and knowing the significance of what you are feeling. Many animals are moved by emotional response patterns, but Dewey claims that only creatures capable of a language that provides a shared reservoir of meanings can actually understand or know what they are feeling. Feelings then "make sense." The difference is between an affective state that is merely had, undergone, and felt, and a state where you understand the meaning of that feeling in your present circumstances. When this happens, we have finally entered the plateau of *mind*, the system of shared meanings arising from social communication, which, for humans, mostly involves language and other forms of symbolic interaction.

The distinction between living things, and living things with mind, is monumental. It opens up a vast system of meaning and thought that allows us to make sense of our situation and ongoing experience, and to respond more intelligently and appropriately to it. Dewey summarizes the critical transition from life to mind as follows:

> Complex and active animals *have*, therefore, feelings which vary abundantly in quality, corresponding to distinctive directions and phases—initiating, mediating, fulfilling or frustrating—of activities, bound up in distinctive connections with environmental affairs. They *have* them, but they do not know they have them. Activity is psycho-physical, but not "mental," that is, not aware of meanings. As life is a character of events in a peculiar condition of organization, and "feeling" is a quality of life-forms marked by complexly mobile and discriminating responses, so "mind" is an added property assumed by a feeling creature, when it reaches that organized interaction with other living creatures which is language, communication. Then the qualities of feeling become significant of objective differences in external things and of episodes past and to come. (Dewey 1925/1981, 198)

This important passage captures the dramatic movement from life, to life with feeling, and finally to life with mind. What was before merely felt—as oppressive, joyful, cautious, anxious, flowing, interrupted—now makes sense when incorporated into a system of body-based meanings that provide a way to understand the *meaning of what you are experiencing*. For instance, what is felt as anxious, uncertain, tense, and cautious becomes *fear*, and perhaps more specifically, fear of public speaking, as one anticipates an upcoming lecture. That uneasy feeling acquires meaning in relation to past experiences (e.g., of previous anxiety-filled attempts at public speaking) and anticipated future attempts.

The result is the fateful emergence of our human ability to mark distinctions, qualities, and actions as meaningful, and therefore as connected to experiences previously undergone and as having implications for projected expectancies of future experiences. Dewey explains, "feelings make sense: as immediate meanings of events and objects, they are sensations. . . . Without language, the qualities of organic action that are feelings are pains, pleasures, odors, colors, noises, tones, only potentially and proleptically. With language they are discriminated and identified" (Dewey 1925/1981, 198).

Dewey argues that it is language, rooted in shared communicative activities, that enables the emergence of mind, with all of its most impressive cognitive operations of conceptualization, reasoning, and evaluative judgments. He tended to regard language as being unique to the human species, but that is an empirical, debatable claim that probably needs to be revised, in light of current animal research (Dunbar 1998; Dunbar & Schultz

2007; McKenna 2013; Tomasello 1999, 2014; Pollick & de Waal 2007). He also tended to regard meaning as dependent on language. However, in the next chapter, we will see that Dewey recognized certain body-based meanings that are not primarily linguistic. Instead, they constitute a vast system of meanings acquired before language, extending beyond language, and giving rise to language. He described this as the *felt sense of a situation as meaningful*, and this is not primarily linguistic. For now, it is enough to see that it is the presence of systems of meanings (whether linguistic or nonlinguistic) that give rise to mind.

Individuals with Minds versus Individual Mind

In Western cultures, most people appear to hold the commonsensical view that each individual person possesses their own unique individualized mind. According to this view, an individual might express what they "have in mind"—what they are thinking—via language that makes this particular thought public and shared. What is inside the mind supposedly gets expressed, that is, *pressed out* into the shared public realm. However, one's inmost feelings and thoughts are essentially *private* and knowable only by the person who is experiencing their thoughts. This is what Daniel Dennett (1991, 1; see also Huebner 2018) has called "the Cartesian theater of mind," in which ideas or concepts supposedly dance on some inner stage of consciousness, observed and observable only by some inner homunculus—the private ego that has exclusive access to its "inner" mental states. What could be more obvious to introspection? You alone know what and how you feel. You alone know your deepest thoughts and feelings. Your personal developmental history creates your specific individual self, with its unique mind, and your unique interiority.

This individual-with-a-mind view, which may seem so obvious to introspection, and which has perhaps a small kernel of truth, is in some ways profoundly misleading. Dewey acknowledged that there are private aspects of conscious experience, but he argued that in Western cultures, especially those dating from the rise of Christianity's doctrine of eternal soul, the tendency to radical individualism and the privatizing of mind has hindered our ability to compose an empirically adequate naturalistic view of mind and self. With the rise of modern science, there was a tendency to de-theologize the Christian notion of the soul as known only by the individual and his or her all-seeing, all-knowing God. This modern version of the self abandoned the theological framework but still preserved the interiority and privacy of mind that was the heart of the Christian view. Descartes's rational ego retains all the trappings of interiority, mystery, and radical freedom, including the belief that the rational ego is not bodily, but rather composed of some mental substance. Kant's Protestant version posits an unknowable transcendent ego (the "transcendental unity of apperception") that is supposed to be radically free and also the ultimate organizing source of our experience of self and world, even though we cannot have any direct experiential access to such a transcendent self, or any knowledge of its existence and nature.

Dewey saw that *mind* is not, in the first instance, individual. It is not a discrete set of processes possessed by an individual and known only to him or her. Instead, mind is shared and public, the result of acquisition of the capacity to engage in cooperative communicative interactions with others, through shared meaning. Mind resides in group cultural transactions, practices, and systems of meanings. Individual people are inextricably embedded and rooted in those shared meanings and values. The "mind" in which individuals grow, and to which they contribute, "is in itself a system of belief, recognitions, and ignorances, of acceptances and rejections, of expectancies and appraisals of meaning which have been instituted under the influence of custom and tradition" (Dewey 1925/1981, 170).

We are born into the world as screaming, cooing, suckling, pooping little creatures, but we couldn't last three days without the continual nurturing care of other people (Stern 1985; Trevarthen 1979). We are profoundly dependent. *Our subjectivity is an intersubjectivity—a being with, in, and through others.* We develop in and through our ongoing intimate transactions with our immediate others. Initially, this is mostly a one-on-one interpersonal give and take, but cultural practices and institutions already enter quite early in our development. Our culture is the soil in which we sprout and from which we are nourished, not just physically, but mentally. We are formed up relative to our common cultural system of meanings, values, practices, hopes, illusions, errors, and successes that make it possible for us to eventually experience our world meaningfully and to engage it intelligently. Mind is a system of ever-developing biases, prejudices, values, and habits, but it is not merely an intellectual and moral straitjacket. Instead, it is also the source of meanings, practices, and habits that support our viability as living creatures and social animals. Consequently, mind both delimits the meanings we bring to our experiences, and it also provides rich resources for all of the creative activities that give rise to new meanings, new understandings, and new forms of engagement, both material and social.

Consequently, mind is not merely a set of fixed channels of thought and action that control every aspect of our being. What Dewey called "individual mind" (as contrasted with "individuals with minds") is the locus of the intersection of inherited habits with individual capacities for creative transformation of experience. The trick is not to fall blindly and passively into fixed templates of cultural customs and meanings, but rather to achieve a delicate balance of cultural embeddedness and critical, innovative remaking of experience. We are inheritors, not radical creators, of our cultural practices, understandings, and values, but we are also capable of innovation, in which our consciousness forms a point of intersection—and redirection—with shared mind.

This view of mind as social and interpersonal, rather than individual, will likely seem counterintuitive to most people enculturated with Western practices, philosophies, and values. Surely, they will insist, there is something going on in "my mind" as I consciously feel, think, and reason. Whatever the source of these interior processes, they are *mine*, and, in the moment that I experience them, they seem to be known only to *me*. As we will see in chapter 5, there is some truth in this received view of individuals having minds. Those who want to reserve the term "mind" only for such conscious events and processes within an individual, are free to do so; however, they are likely to overlook most of what goes into human meaning and thought. In particular, they will exclude and overlook vast nonconscious operations of meaning and thought, and also all of the interpersonal and cultural dimensions of the self. We are so habituated to thinking of mind as something possessed by individual persons that perhaps it would be easier for us if we could substitute "culture" or "cultural meaning" for Dewey's use of the term "mind," in order to indicate all of the systems of meaning, practices, and values shared by a people and constituting the source of the meanings that are available to individual subjects. Dewey would be fine with that substitution, but he would rightly point out that, whatever name we choose, we will still always need an account of how the individual is nurtured within, and draws its meanings, thoughts, values, and practices from, something which is in some sense external to, and wider in scope than, individual mental processes, while also making possible those very same internal thoughts.

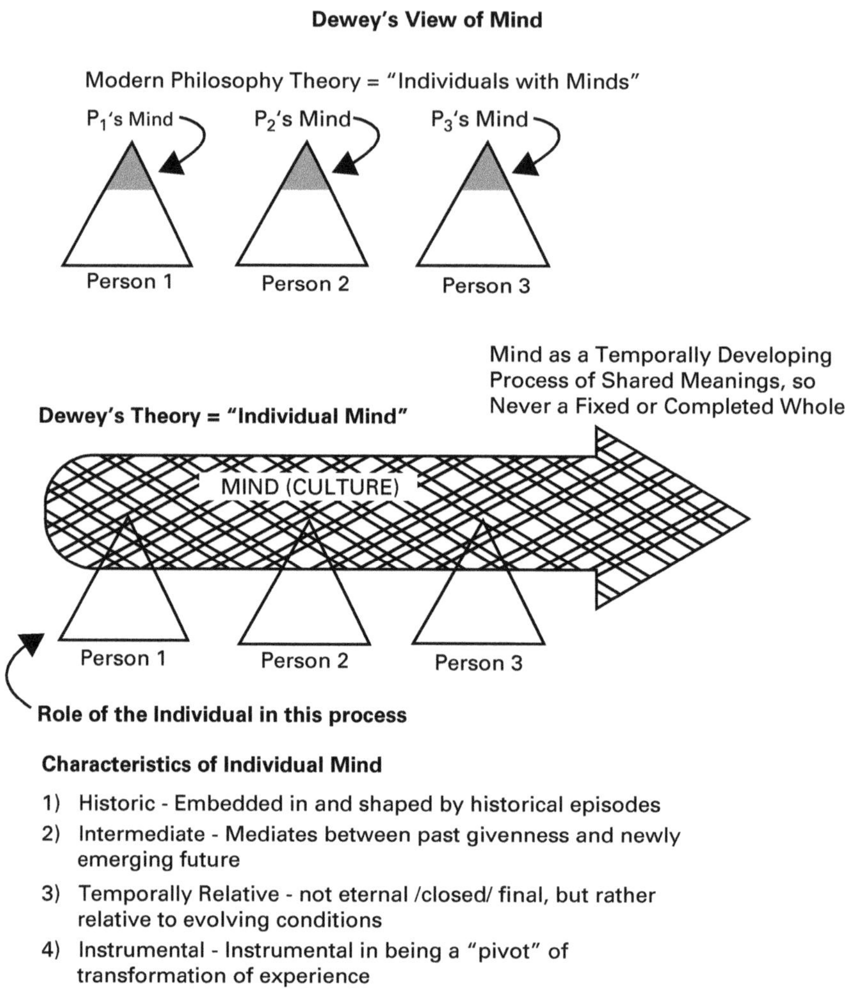

Figure 3.1
Dewey's model of mind and culture.

Dewey, and George Herbert Mead (1928/1972, 1934), appreciated the social constitution of the self (see chapter 7), long before that idea became the bedrock of social and developmental psychology in the second half of the twentieth century. Earlier, Fichte (1796/2000) had seen the interpersonal nature of selfhood (as intersubjectivity), and Hegel (1807/2018) saw how the individual was embedded in, and nurtured by, an ever-developing system of cultural resources and achievements that he called objective spirit. Dewey basically naturalized objective spirit and called it mind.

Dewey also anticipated aspects of what today is known as "extended mind" (Clark & Chalmers 1998), which emphasizes all of the ways our concepts, memory, calculations, and thinking get off-loaded onto our environment. This off-loading began with sticks and stones (for counting and measuring) and clay tablets and pots inscribed with figures, later to be replaced with paper and pencil, and now extended to smartphones, computers, and other sophisticated information technologies. This process has been profoundly expanded by means of new communicative systems and modes of information storage and transmission, such as Wikipedia, Facebook, and Twitter. These recent developments have made it far less strange to think of mind as extending far beyond the confines of skin and skull.

The key insight in all of this is that "No man is an island, entire of itself/ Every man is a piece of the continent, a part of the main" (Donne 1624). The myth of the individual, independent, autonomous self is a dangerous illusion—dangerous because it pretends that we are free agents of our self-creation, unconstrained by our social, cultural, and historical situatedness. Dewey argues that when individuality "is conceived of as original, eternal and absolute" we fail to see that it is actually "historic, intermediate, temporally relative and instrumental" (Dewey 1925/1981, 174). Individual selfhood is *historic* in being embedded in and shaped by historical events. It is *intermediate,* insofar as it mediates between past givenness and a newly emerging future. It is *temporally relative,* as developing over a particular temporal span, thereby inheriting prior values, institutions, and practices which it may actively engage and transform. The self is not eternal, closed, and fixed, but rather is relative to evolving conditions. And it is *instrumental* as a means of engendering new meanings and reconstructing culture.

Decades before the discovery of what is now called "the enactive self," Dewey realized that, rather than being a fixed structure, the self is an ongoing enactment of meaning, activities, and values. Thompson (2007) and Varela et al. (1991) first characterized *enaction* as the view that "(1) perception consists in perceptually guided action and (2) cognitive structures emerge from the recurrent sensorimotor patterns that enable action to be perceptually guided" (Varela et al. 1991, 173). Varela et al., in turn, drew on Merleau-Ponty's emphasis on activity as the basis for all experience and thought: "since all the stimulations which the organism receives have in turn been possible only by its preceding movements, which have culminated in exposing the receptor organ to external influences, one could also say that behavior is the first cause of all the stimulations" (Merleau-Ponty

1963, 13). Dewey's version of this is his account of the initiating role of action in all perception (Dewey 1896). Gallagher (2016) summarizes this enactive, action-based generation of meaning: "the brain—attuned by evolutionary pressures and by the individual's personal (social and cultural) experiences, in its interwoven relations with body and environment—or in other words, the system as a whole, with its own structural features that enable specific perception-action loops, that in turn shape the structure and functioning of the nervous system—*responds to* the world rather than *represents* it" (Gallagher 2016, 364; see also Heelan & Schulkin 1998).

Dewey thought that language was the supreme tool (the tool of all tools) for that ever-developing system of meanings, practices, and values that he called mind, or culture. He held that it is our remarkable human capacity to experience, make, and communicate meanings that makes culture possible. The plateau of mind arises only with language, which supposedly separates humans from all other animals. Therefore, we now need to look more deeply into the nature of language, and especially into the nature of meaning, if we hope to understand how mind, thought, and knowledge are possible. Is all meaning linguistic, or is language dependent on meanings and modes of symbolic interaction that operate mostly beneath the level of consciousness?

4 Meaning and Thought

In one of the all-time great openings to a chapter in a philosophy book, Dewey writes: "Of all affairs, communication is the most wonderful. That things should be able to pass from the plane of external pushing and pulling to that of revealing themselves to man, and thereby to themselves; and that the fruit of communication should be participation, sharing, is a wonder by the side of which transubstantiation pales" (Dewey 1925/1981, 132). Transubstantiation—the transformation of the bread and wine of the Eucharist into the body and blood of Christ—would indeed be miraculous, but, says Dewey, if you want something that is *really* miraculous, consider human communication, the use of signs to utilize shared meaning to further cooperative interactions! However awe-inspiring the glories of language might be, the challenge for a naturalistic philosophy is to explain these glories of human meaning, language, and communication, not as miraculous occurrences, but as wholly natural emergent functions.

In the previous two chapters, we described an emergentist metaphysics, in which there is a continuity of development and structure from the plateau of inanimate matter, to living beings, to creatures with mind. Dewey regarded the development of language as the key to the emergence of mind—the realm of shared meaning that is culture. Now we need to delve more deeply into the nature of meaning, especially since meaning and conceptualization are often the strongholds of disembodied views of mind and thought. There is a long history of treating concepts as abstract entities populating an inner theater of mind. Moreover, even among philosophers who equate mind with brain, there is still too often a tendency to think of concepts as processed in a different part of the brain than those areas involved in perception and movement. The result is a view that keeps perception and conception apart, as if they were radically different in kind. In contrast, the naturalism we are developing sees all meaning as rooted in sensory, motor, and affective processes, and it treats concepts as continuous with, and constituted by, sensory–motor structures and processes. This is what we will call embodied meaning.

Meaning as Experience

Consider a fairly simple primitive animal that we will call "Little Critter," that picks up on stimuli in its surroundings and selectively responds to them as they afford nutrition, water, or shelter. Little Critter might have a tiny brain, or no brain at all (having only a primitive nervous system), but it nonetheless finds ways to remain viable in certain environments. It detects nutrients and moves to ingest them. It turns away from harmful substances. From our third-person vantage point, we might anthropomorphically project onto this creature needs, desires, and planning. However, although Little Critter has values and responds to stimuli via capacities resulting from the evolutionary development of its species, it has no awareness or feeling of needs, wants, or values. In the sense of meaning that is available to humans, nothing *means* anything to our Little Critter, even though it clearly responds differentially to its surroundings, with avoidance and approach responses, preferentially seeking some things and avoiding others.

Maybe the good news for Little Critter is that he doesn't lie around worrying about his finitude, about how his girlfriend ignored him last night, about the results of his upcoming medical examination, or, frankly, about *anything*! We humans will tend to think that Little Critter is missing out on something that is very important, like the ability to assess the meaning of what is happening to him, to problem solve, to plan and execute an intelligent response, and to imagine remarkable future possibilities of consummatory action. In Dewey's terms, Little Critter is a living being, possessing some (though maybe not all) of the characteristic capacities of many animals, but he never rises to the level of mind, or culture. From the human perspective, what he's missing out on is massively important, making all the difference between life as determined only by routinized habits, and a life capable of evaluation of the meaning of a situation, and of intelligently problem solving, as a basis for the creative transformation of experience.

The transition from the physical realm to the psychophysical and on up to mind can be understood as a process by which meaning emerges in animal life, via the increasing complexity of the organism and its environment, and then reaches a particular consummation in human mind. As we will see, Dewey typically locates meaning at the plateau of mind, where signs can stand for certain experiences, past and future. However, Dewey also recognizes the emergence of a kind of meaning even for nonhuman animals for whom qualities become significant:

> Qualities actually become specifically effective however, in psycho-physical situations. Where animal susceptibility exists, a red or an odor or sound may instigate a determinate mode of action; it has selective power in maintenance of a certain pattern of energy-organization. So striking is this fact that we might even define the difference between an inanimate body and a vital and psycho-physical one, by saying that the latter responds to qualities while the former does not. In this response, qualities become productive of results, and hence potentially significant. That is, in achieving effects they become connected with consequences, and hence capable of meaning, knowable if not known. (Dewey 1925/1981, 205)

Although Dewey reserves the plateau of mind for the realization of meaning in its richest sense, here he recognizes that a continuous development from life to mind requires at least a primitive notion of meaning operative on the plateau of life, even prior to the emergence of communally shared symbol systems, or signs. When qualities become significant for an animal, they are meaningful as affordances for that creature.

This thrusts the question of meaning front and center. We have to investigate what meaning is, where it comes from, and why it matters so much for our lives. We begin with the following general definition: *The meaning of any object, quality, or event is the experiences (past, present, and future) it evokes or enacts for a person.* One commentator has observed that in chapter 5 of *Experience and Nature*, entitled "Nature, Communication, and Meaning," there are over twenty different statements of the form "meaning is . . ." Although we won't take the time to work through these many statements, we submit that they are all either versions or implications of this general definition.

The meaning of something is not any sort of abstract entity or nugget of essential features entertained and manipulated within a private, interior realm of mind. Rather, the meaning of something is what it does in, through, and to experience. The meaning of something is its implications for experience and action, where those implications can be either conditions of a particular object or event, involving its previous history, or also the expectancies for future possible experiences that it affords. Dewey gives an example of the meaning of fire: "The ultimate meaning, or essence, denominated fire, is the consequences of certain natural events within the scheme of human activities, in the experience of social intercourse, the hearth and domestic alter, shared comfort, working of metals, rapid transit, and other such affairs" (Dewey 1925/1981, 150). The meaning of fire is what it does to and for us, including how it affects us, and also what we can do with it by way of altering experience. The meaning of something past is remembered experiences that are carried forward, to shape present meaning, and then to provide expectancies about future possible events and experiences. With respect to fire, as the saying goes, "Once burned, twice shy." We learn part of the meaning of fire by its ability to burn things. We learn more of its meaning as we discover how we can use it to effect certain changes. And we learn even more of its meaning in the roles it plays in various practices and rituals of our lives. Since the affordances of different objects and events may change over time, relative to our interests and values, meaning, although often relatively stable within a culture, is never fixed, complete, or unchanging. As conditions change, and new possibilities for action and cooperation arise, new meanings emerge. The meaning of "fire," for example, expands, as we discover new affordances of fire and new possibilities for working with it to effect changes in the world.

Dewey's account of meaning is inextricably interwoven with his account of language and communication. The operative idea is that the development of language profoundly extends and enriches possibilities for communication and cooperative activity, by providing the conditions for meaning, which is shared, public, and transactive. Dewey illustrates this communicative process with his example of one person (*A*) requesting another (*B*) to bring him a flower. Person *A* points to the flower as he speaks. *B* responds, not just to the pointing, or to the object pointed to, as such, but rather to the complex transaction going on between *A* and *B*, as an indication of an intent on *A*'s part to get a certain cooperative response from *B*, namely, bringing *A* the flower. Dewey concludes: "He (*B*) perceives the thing as it may function in *A*'s experience, instead of just ego-centrically. Similarly, *A* in making the request conceives the thing not only in its direct relationship to himself, but as a thing capable of being grasped and handled by *B*. He sees the thing as it may function in *B*'s experience. Such is the essence and import of communication, signs and meaning. Something is literally made common in at least two different centers of behavior" (Dewey 1925/1981, 141).

This example highlights how meaning arises in cooperative communicative behaviors and transactions, based on awareness of shared intentions, and drawing on culturally shared values, interests, and practices. However, we will argue that we need not limit meaning only to explicit communicative transactions with other people, because once a system of meaning is developed (and continues to develop over time), things can become meaningful to individuals, even when they are not engaging in cooperative interpersonal exchanges. For example, a tree can be meaningful to me via my actual and potential interactions with it, even when I am not communicating with another person about it. The tree can be a symbol of life, a shelter from a rainstorm, a structure for a child to climb, a source of lumber, a remover of carbon dioxide and a creator of oxygen, a provider of shade on a sunny day, and much more. These meanings implicate mind—the developing system of shared meanings—but experiencing a tree as meaningful in some of these ways need not always immediately require communication with others. Thus, Dewey's flower example is a prototypical instance of meaning exchange, but it does not represent the full of range of situations in which meaning arises. We therefore need to explore how meaning arises through our engagement with our environment, via bodily processes of perception, feeling, and action, and how it is then assimilated into cultural systems of meaning.

Embodied Meaning Arising from the Qualitative Unity of a Situation

The field known as philosophy of language was created and developed chiefly in the second half of the twentieth century, and mostly in so-called analytic departments of philosophy. It was based on the assumption that all meaning, thought, and knowledge are mediated by language. Richard Rorty (1967) adopted Gustav Bergmann's (1960) phrase "the linguistic turn" as the title of Rorty's influential anthology of writings by prominent philosophers who saw language as the central focus of a newly emerging "analytic" or "linguistic" philosophical orientation. The guiding theme was that, since all cognition is supposedly based on, and mediated by, language, the analysis of linguistic concepts and propositions is the key to understanding mind, thought, and knowledge. If all concepts are language dependent, then, as John Searle claimed, with his Principle of Expressibility, "whatever can be meant, can be said" (Searle 1969, 19). Linguistic analysis was regarded as the way into our concepts, reasoning, and knowledge. Logical empiricism proposed that cognitive meaning and thought could be analyzed into a logical component, specifying a set of universal logical relations, and also into an empirical component, where concepts and propositions were thought to correspond to mind-independent states of affairs. Thus, you got concepts allegedly grounded in experience, and then organized and related by logical principles, issuing in determinate meanings, propositions, and knowledge claims.

The idea that all meaning and thought are linguistic in nature has been dismantled by empirical cognitive science and cognitive linguistics (Feldman 2006; Johnson 1987; Lakoff 1987; Langacker 1987–91) over the last three or four decades. Even before respected logical empiricists like Hempel, Carnap, and Quine began to detect fatal flaws in their own logical empiricist program, Dewey had already catalogued the fundamental errors resulting from the view that all experience, meaning, and thought are mediated by language, understood as words and sentences used in the performance of various speech acts. He saw how treating all experience and cognition as linguistic leads us to overlook the body-based meaning that exists prelinguistically (in children) and then, in adults, constitutes a vast realm of embodied meaning that extends beneath and beyond language, while giving rise to language. Empirical research from the embodied cognition perspective supports, extends, and enriches Dewey's account. Dewey was a progenitor of many of the major themes of embodied cognition theory (Crippen & Schulkin 2020; Johnson 2017, 2018).

Dewey's account begins with the fact that each of us, at any moment, finds ourself entangled in an experiential situation from which all our meaning arises, and through which it is modified. In his important, but not widely known, essay "Qualitative Thought" (1930), Dewey explores how relevant, organized thought is shaped by the patterns of organism–environment interaction within a qualitatively unified situation: "The term 'situation' in this connection is signified the fact that the subject-matter ultimately referred to in existential propositions is a complex existence that is held together in spite of its internal complexity by the fact that it is dominated and characterized throughout by a single quality. By 'object' is meant some element in the complex whole that is defined in abstraction from the whole of which it is a distinction" (Dewey 1930a/1988, 246). Within the context of a particular meaningful situation, we then select objects and events as the focus of our attention, based on our habits of perception and thought, our needs, and our values. The point of insisting that we begin, experientially, not with particular objects, but rather with a whole situation unified by a pervasive quality, is that consideration of what is relevant in any process of thought is determined by the experiential setting in which we find ourselves at a particular point in time. That circumscription of relevant objects and events will seldom be clear and distinct, since which factors are relevant will tend to vary given one's situation, acquired habits, interests, and values. As Merleau-Ponty observed, surrounding focal objects of attention, there will always be an expanse of possible meanings fading off into an indefinite horizon of significance (Merleau-Ponty 1962, 1963). Merleau-Ponty identified this center-periphery structure as present not just in visual perception (where there is a focal point within a broader visual horizon), but as the horizonal structure of cognition in general, where there are both focal and peripheral meanings. As we move through the world, shifting our attention from one thing or event to another, the horizon of possible meanings shifts, so that what was central and focal might become peripheral, while something previously on the horizon now becomes focally important. New meaning emerges.

Dewey saw that logic alone will not tell you what is relevant to your present line of thought. For example, knowing *modus ponens* (namely, that "if *a* then *b*, *a*, therefore *b*") does not tell you what the pertinent *a* and *b* are, nor why they are relevant to each other and to the context of your present inquiry. The logic cannot specify what you are thinking about. *That*, the relevance, is determined by the specific situation in which you find yourself. Moreover, that whole situation must be somewhat circumscribed, in order to limit the relevant factors in a train of thought (Pappas 2016)

Logic is thus not a system of absolute, pure, and universal relations that are supposedly realized in thought. Rather, logic emerges in and from processes of empirical, experiential inquiry (Dewey 1938/1991). It is embodied and situationally embedded. Logic doesn't generate relevance, because those similarities and contiguities that we perceive depend on prior selection of what counts as relevant objects, properties, and relations in an experiential state of affairs. Things have to be felt as contiguous and/or similar before they can be known as such, and reasoned about. That togetherness is revealed in the experienced situation, and the meaning of anything is determined only in its full, felt experienced context.

Commonsensically, we think of objects as existing in a mind-independent world, having certain inherent properties, and standing in objective relations to other objects. We think of ourselves as learning about the world by associating objects (and events) based on their properties and relations. David Hume (1739) argued that learning occurs through associations based on contiguity (i.e., togetherness in space and/or time) and resemblance (i.e., having shared similar properties or relations). This enlightenment associationist psychology takes contiguity and similarity to be objective givens, and then asks how they generate the associations we make as we learn from experience.

Dewey doesn't deny that we learn by association. In fact, he affirms that "thinking as an existential process is all one with controlled association" (1930a/1988, 260), but he argues that the associations we make are controlled by the qualitative unity of the situation, and are not mind-independent relational givens. Contiguity and resemblance are thus determined by the situation. His example of contiguity is our association of nests and birds. The traditional view would be that we associate nests and birds because we see them together over and over again, with each occurrence of togetherness strengthening the association, thereby making it more and more likely that we will think of birds with nests together. But notice how often we see nests without birds, and even more often birds without nests? Also, we see nests along with leaves, twigs, and branches more than with birds, so why is the bird–nest association primary? What grounds the association is, rather, the unified experience of a-nest-with-a-bird-in-it, as a functional whole—birds build nests and live in them. Dewey concludes that:

> The association of bare contiguities is a myth. There is an indefinite number of particulars contiguous to one another in space and time. When I think of a nest, why does a bird come to my mind? As a matter of contiguity, there are multitudinous leaves and twigs which are more frequently and more obviously juxtaposed than is a bird. . . . Such questions suggest, I hope, that even in seemingly casual cases of association, there is an underlying quality which operates to control the connection of objects thought of. It takes something else than contiguity to effect association; there must be relevancy of both ideas to a situation defined by unity of quality. (Dewey 1930a/1988, 258)

Besides contiguity, the other main proposed basis for association has been similarities (i.e., resemblance). Dewey criticizes F. H. Bradley's argument that certain forms and patterns in nature are objectively similar, and that those similarities ground our knowledge of things. Bradley argues (Dewey 1930a/1988) that we see two promontories, one in England and the other in Wales, as similar by virtue of shared formal relations which are the same in the two different objects. Dewey rejects this account by observing that the form of something is not an isolated element in a scene, but rather "an arrangement of patterns and elements" within a whole (1930a/1988, 259). A gently curving line in a Matisse painting is not meaningful in itself, nor does it have the same meaning anywhere it occurs, but rather its meaning arises in relation to other forms and lines within the pervasive unifying quality of that particular painting. Dewey concludes that both contiguity and resemblance are identified only within the qualitative unity of the whole situation: "The only way that form or pattern can operate as an immediate link is by the mode of a directly experienced *quality*, something present and prior to and independent of all reflective analysis, something of the same nature which controls artistic constructions" (Dewey 1930a/1988, 259).

Dewey is making a frontal assault on the received version of associationism, according to which thought is a process of taking independent, pregiven forms, patterns, or characteristics and then associating them because of some alleged objective relation they share (such as contiguity or resemblance). Dewey's alternative is that "the quality of a situation as a whole operates to produce a functional connection" (1930a/1988, 257), in this case, perceived contiguity or resemblance. The determination of what is relevant to a course of thinking depends on the whole complex situation resulting from the interaction of a person and their environment. Dewey's account of relevance is a precursor to the view that what matters is the predictive contexts of the contiguity and resemblance relations, which is where modern associationist learning theory was led (Dickinson 1980; Miller 1959; Rescorla & Wagner 1972).

The unity of a situation is *qualitative*, and it is *felt*, as much as it is *thought*. However, the fact that it is felt does not make it merely subjective. The unity is that *of a situation*—an objective state of affairs, relative to the habits, needs, and values of the organism engaging its surroundings. We feel a meaningful unity (as things "make sense to us"), but that unity is the objective unity of a whole situation, prior to any selection of objects, properties, and relations that become the focus of subsequent reflective thought.

The qualitative unity is not a distinct isolatable quality of some object or event. It is what unifies and directs thought within a situation: "The situation as such is not and cannot be stated or made explicit. It is taken for granted, 'understood,' or implicit in all propositional symbolization. It forms the universe of discourse of whatever is expressly stated or of what appears as a term in a proposition. The situation cannot present itself as an element in a proposition any more than a universe of discourse can appear as a member of discourse within that universe" (Dewey 1930a/1988, 247). The fact that this qualitative unity cannot be made explicit suggests why the idea of a pervasive quality (and its crucial role in all thought) has been overlooked, or even rejected, by most philosophers. If it can't be described as a discrete, individual quality, then what can one possibly say about it?

Two Neural Systems Underlying the Pervasive Unifying Qualities

Dewey's fundamental idea of the crucial role of pervasive qualities in meaning and thought has not been well received either by analytic philosophers or by phenomenologists. Why not? The main reason is that, as noted earlier, the pervasive quality cannot be adequately described, because then it would be treated as a particular quality, and not a unifying quality of a situation. The second reason is the tendency to focus exclusively on the quality as felt, and then to regard that as entirely subjective, whereas the qualitative unity is supposed to be an objective character of a situation. The third reason is that something as expansive as a pervasive quality would seem to be so vague and ill-defined as to be useless for analytical purposes.

Dewey's account of such a quality focuses on its necessary role in thinking. It is the qualitatively unified situation that roots thought in experience and determines the relevance of objects and events. One is left wanting more—more argument and more supporting evidence. One source of evidence unavailable in Dewey's day that strengthens the argument for the existence and importance of the qualitative unity of a situation comes from the neuroscience of motive control mechanisms that underlie both motor control and thought.

The neuroanatomical details are daunting, quite complex, and in some spots speculative (Chao et al. 2002; Haxby et al. 2001; Martin 1998, 2007; Martin et al. 1996; Ungerleider & Mishkin 1982). We previously described how animals routinely engage the need–search–satisfaction pattern as they pass in and out of sync with their environment. The goal of their appetitive search is the maintenance and, after disruption, the restoration of homeostatic and allostatic equilibrium. All of this bodily monitoring and action is thus shaped by values, capabilities, and resources tied to the organism's viability in its environment.

The sensory and motor processes involved in this quest for viability are rooted in two prominent systems in the brain. First, based on habits of perception and action developed via past experience, animals project expectancies about what they will encounter at the sensory–motor interface with their surroundings. If these predictions are mostly satisfied in experience, the animal moves forward under the impetus of impulses and engages its world more or less automatically and unreflectively. This feeling of what William James (1890) called "furtherances" is a sense of relative elation and an openness to the world. The self seems at home in its surroundings, and moves forward more or less fluidly.

However, when discrepancies arise between projected expectations and what is experienced, one's situation becomes indeterminate and uncertain. We feel what James (1890) called "hindrances" to action. We become anxious and unsure about what to think and do. What ensues is either continued confusion and inability to take action, coupled with growing anxiety, or else the search process that Dewey described as a form of problem solving. We attempt to recalibrate our expectancies as we consider possible ways of transforming our situation through thought and action, in order to accommodate new conditions.

There are two pathways (dorsal and ventral) in the brain for these two motive control systems. Via a specific neural architecture that we will not describe here (see Johnson & Tucker 2021; Rauschecker & Scott 2009; Tucker & Luu 2012; Ungerleider & Miskin 1982), the dorsal stream creates a meaningful, affect-rich, and relatively fleeting contextual representation of a situation. The ventral stream provides a sustained focus on an object. This dorsal/ventral distinction was originally made in the exclusive context of vision, where the difference was between spatial location vision and object vision (Ungerleider & Mishkin 1982). The dorsal stream is tied to the occipital/parietal regions of the brain, whereas the ventral stream is tied to occipital/temporal regions. Both regions are informed by extrastriate cortex that provides feedback from the sensory–motor interface with the environment (Tucker & Luu 2012). Both are critical for visually guided behavior, involving attention to an object (ventral stream) at some spatial location (dorsal stream).

For humans, vision dominates neocortical expression, and constitutes about a third of total brain in mass (see Koch & Reid 2012). This anatomical basis for perception of an object situated in space, along with the associated affordances of expected relations, is critical for the organization of action. Later this distinction was suggested as relevant to the perception of objects versus organized control toward the objects, and, of course, both are running continuously in everyday experiences (Milner & Goodale 1995). Similar distinctions have been made for other sensory systems (e.g., audition, Rauschecker & Scott 2009). The theme—the relation of objects to their spatial context—is general across probably all sensory modalities, but its intellectual roots are within the study of vision, and that should not be surprising, given the importance of vision in navigating within an environment.

In general, the dorsal stream, which provides a sense of context, is *phasic*, meaning that it rapidly habituates and quickly passes, unless it is stimulated by new experience. It has been called the *Where* pathway.

In the presence of an indeterminate situation, the ventral track, by contrast, selects focal objects of attention within the broader context supplied by the dorsal track. This has been called the *What* pathway. The detailed focus on an object is *tonic*, meaning that it sustains over time, due to sensitization to the object. This allows for a more focused, detailed, and sustained attention, which makes possible the projection of revised expectancies for experience. In ordinary experience, the working together of the ventral and dorsal pathways results in attention to an object of interest, situated within a spatial setting, as perceptual context. The attentional focus (Desimone 1996; James 1890; Pashler 1998) captures the object, and holds it in focus, as we explore and forage for further meaning.

Of course, in real time adjudications these two streams are not totally separate. They interact effortlessly to situate the object in context and to determine its meaning for a prepared brain, controlling a body that is acting within suitable environments.

We suggest that when Dewey described the pervasive unifying quality, he was describing the mostly unconscious operation of the dorsal pathway that projects a holistic, affect-rich context (Johnson & Tucker 2021). Then the ventral pathway allows focused attention on an object, to discern its meaning in the present context (Emery 2000; Perrett & Emery 1994; Rolls 1996, 2000; Rolls & Treves 1998). This is the mechanism for engaging specific objects with discrimination and analytic focus.

The dorsal pathway thus provides a plausible basis for Dewey's qualitative unity of a situation. The whole situation is first felt as significant, relative to our visceral needs, and it then determines what is relevant (as a projected set of expectancies) for sustained, detailed focus on an object, as we continuously adapt and cope with our diverse needs and situations. Other investigators have suggested a similar distinction (Gibson 1979; Neisser 1994). J. J. Gibson, for instance, emphasized the importance for the organization of action of where objects are, what their affordances are, and what their value and expected outcomes are, relative to predictive coherence.

Neural systems involved in actions can be active whether we move or not (Jeannerod 1997, 1999). For example, imagining an action, such as reaching for a cup, activates many of the same regions in motor and premotor cortex as if one were actually reaching for it. Similarly, imagining a visual scene activates many of the same visual cortical regions as if the objects were before one (Kosslyn 1980, 1994). The same holds for audition (Calvert et al. 1997). All of this predictive activity gives rise to expectations rich with meaning.

Embodied Meaning as Sense and Signification

The next step in a naturalistic account of meaning and thought is to determine how anything begins to be meaningful for us within a particular context. How does meaning emerge, first via nonconscious processes, and then eventually become available for conscious reflection and revision? Dewey identified two fundamental processes of meaning-making, which he called *sense* and *signification*. *Sense* is the feeling of the appropriate situation as meaningful and understandable, while *signification* is the use of signs to indicate meaning (as evoked experiences) of particular objects and events within that situation.

Experience starts with a situation involving the interaction of an organism with its environment. A particular situation "makes sense" to us because it is marked by a pervasive qualitative unity that determines which phenomena are experienced as relevant within a given setting. The sense of a situation involves the two notions of feelings that we described in chapter 3; namely, nonconscious discriminations (i.e., sensitivity) and also those consciously felt changes in our body-state as we engage aspects of our environment (i.e., conscious feelings). Normally, when we speak of feelings, we are thinking of this latter consciousness of our changing body-states.

This commonsense notion of conscious feelings gives rise to adjectives and adverbs such as sweet, tart, stinky, cold, warm, bright, high-pitched, rapid, slow, meandering, rushing, halting. Such terms indicate the conscious feeling of certain qualities and patterns of sentient experience that are a fundamental part of sense.

However, this awareness of feelings does not indicate what these are qualities *of*. Dewey therefore reminds us that we can *have* a feeling, without *knowing* what we are feeling. *Sense is the first step in emerging from feelings had to feelings known as significant for experience. It sets the whole context within which signification then selects objects and events with meaning.* Dewey summarizes the sense versus signification distinction as follows:

> Sense is distinct from feeling, for it has a recognized reference; it is the qualitative characteristic of something, not just a submerged unidentified quality or tone. Sense is also different from signification. The latter involves use of a quality as a sign or index of something else, as when the red of a light signifies danger, and the need of bringing a moving locomotive to a stop. The sense of a thing, on the other hand, is an immediate and immanent meaning; it is meaning which is itself felt or directly had. When we are baffled by perplexing conditions, and finally hit upon a clew [*sic*], and everything falls into place, the whole thing suddenly, as we say, "makes sense." . . . But the meaning of the *whole* situation as apprehended is sense. (Dewey 1925/1981, 200)

A naturalistic theory of mind, thought, and language cannot just posit a mental repository of concepts and propositions as it pleases. It has to explain how concepts and propositions emerge from sensory, motor, and affective processes—from the bottom up, so to speak. Sense provides the meaningful context within which signification discriminates objects, properties, and relations. Sense, unlike signification, does not rely primarily on sign systems connected to language. It is immanent meaning arising as a feeling of the qualitative unity of one's situation and one's sense of being able to orient meaningfully in that situation. It is the beginning of knowing what one is feeling and employing that meaning in action.

Relative to the sense of a given situation, signification then gives us the meaning of an object or event in terms of the experiences (past, present, and future) that it evokes and enacts. A sign (i.e., an image, word, or object) is used as an indication of past experiences with some object or event, and also as an anticipation of possible experiences to come. This is what we typically think of when we speak of the meaning of a word.

In contrast to the objectivist view of concepts as rigidly demarcated and defined by a set of necessary and sufficient conditions (i.e., a set of essence-defining features), Dewey's pragmatist account recognizes the internal complexity of most concepts. Consider, for example, the meaning of "water." Water is that clear, wet, flowing substance that falls down from the sky, collects in rivers and lakes, turns into snow and ice when it gets cold, melts back into liquid form when it warms up, and evaporates into an invisible vapor when the sun shines hot. It is what we wash our clothes and bodies in, what we pour on plants to make them grow, what we swim in, what we drink to slake our thirst, what we use in baptism rituals to cleanse the soul, and what we implore the gods to send us in drought, to save our crops. All of these experiences are realized neurally in activated functional clusters of neurons in the sensory, motor, and affective areas of our brains (Barsalou 1999; Martin 2007). Water is not just one thing, but a cluster of interrelated phenomena, relative to our values, needs, and cultural practices.

A concept like *water* is multimodal, which means that it involves activation of multiple sensory and motor systems. When we read or hear the term "water," areas of our brains are activated that process our experience of seeing, touching, tasting, and smelling what we call water. The meaning of "water" is partly the sensory, motor, and affective experience it calls up for us, and these experiences are based on activation of functional neural clusters in the various sensory and motor regions of the brain. However, the meaning of water is not just a set of perceptual experiences. It also includes the things we *do* with water, such as the aforementioned actions of watering plants, washing things, mixing cement, squirting children with a hose, performing baptismal rituals, and swimming in a cool mountain lake.

The meaning of water, and typically of the vast majority of our concepts, is thus complex, multimodal, and functionally determined. Regions of the brain within temporal cortex, such as the amygdala, are involved in capturing the "gist" of things; a rough, good enough detail in order for us to function effectively (Adolphs et al. 2001). Getting the gist of things, the sense of things, involves understanding the sense of the relevant context (spatial or semantic) and also signification of objects salient within that context, through use of signs. Both sense and signification are thus fundamental to the coherence of action, as we forage both to survive and to understand our environment. Inherent in this is capturing the meaning of things and events that matter, or should matter, and this capability of finding what is both relevant and significant for our present situation has been a key factor in the evolution of our expanding capabilities, such as our capacity for self-corrective inquiry.

Another major issue is the relation of meaning and language. Is all meaning dependent on language use, or is there prelinguistic meaning, too? Dewey claims that it was the development of language that made possible the remarkable human capacity for shared meaning through cooperative communicative activity. When a quality is no longer taken merely as something immediately experienced, but instead becomes a sign of something distant in time or space, a new world of meaning is opened up. Things are not just what they immediately are, but point beyond themselves to possible meanings, or affordances of experience:

> Brute efficiencies and inarticulate consummations as soon as they can be spoken of are liberated from local and accidental contexts. . . . Events when once they are named lead an independent and double life. In addition to their original existence, they are subject to ideal experimentation: their meanings may be infinitely combined and re-arranged in imagination, and the outcome of this inner experimentation—which is thought—may issue forth in interaction with crude or raw events. . . . Where communication exists, things in acquiring meaning, thereby acquire representatives, surrogates, signs and implications, which are infinitely more amenable to management, more permanent and more accommodating, than events in their first estate. (Dewey 1925/1981, 132)

It is in this sense that the emergence of systems of meaning—Dewey's *signification*—is co-extensive with the rise of linguistic communication. The relation of meaning to language is not causal, but rather a co-emergence and co-constitution.

However, if meaning is supposedly dependent on cooperative use of language, this raises a fundamental question about the status of sense. In the passage on sense versus signification quoted earlier, sense—as what Dewey called "immanent meaning"—does not seem to rely primarily on language. So, does all meaning really depend on language?

We have been suggesting that the sense of a situation is not, for the most part, dependent on language, but rather underlies and gives rise to linguistic communication, as signification. Prelinguistic children are already beginning to be able to "make sense" of a particular situation, and it is thereby meaningful to them, though not in any way they can articulate conceptually and linguistically. The sense they experience is certainly not as rich and expansive as their older peers, and not even close to that of most adults, but it *is meaningful*, and becoming more meaningful day by day, as they learn to organize and understand their blossoming experience within their developing system of meanings. Dewey thus emphasizes the importance of the use of signs (i.e., signification) as making possible vastly expanded and enriched communicable meanings. However, none of this would be possible without the prior existence of a situation that "makes sense." The process by which some object or quality becomes meaningful is not an inexplicable eruption of new meaning, but rather a gradual accumulation of the sense of a particular situation, which, in turn, gives rise to the enrichment of meaning through sign systems. Dewey focuses on culturally shaped communicative practices for sharing meaning and coordinating activity, because they constitute the consummation of our attempts to discern the meaning of things, but he insists that none of this high-flying achievement would be possible without our body-based sense-making.

Dewey's emphasis on meaning as communicative interaction using language, raises the related question of the status of nonlinguistic forms of symbolic interaction, such as dance, architecture, spontaneous gesture, theater, music, and ritual practice. We have been suggesting that all of these activities are full-blown enactments of meaning, even though they all involve meaning that is not always dependent on language. One way to approach this issue is to say that all meaning is linguistic, but then to add that by language we mean something more than just spoken or written words—something that includes all of those activities (gesture, dance, architecture, painting, etc.) as part of meaning. Indeed, Dewey explicitly understands language in this more expansive sense: "Language is taken in its widest sense, a sense wider than oral and written speech. It includes the latter. But it includes also not only gestures but rites, ceremonies, monuments and the products of industrial and fine arts. A tool or machine, for example, is not simply a simple or complex physical object having its own physical properties and effects, but is also a mode of language. For it *says* something, to those who understand it, about operations of use and their consequences" (Dewey 1938/1991, 51–52). On this reading, all meaning is linguistic, but language is not restricted to this narrow sense of use of words since it includes many nonlinguistic forms of symbolic interaction.

A second interpretation is to say that sense and signification represent two different dimensions of meaning, with the former not primarily dependent on words (language), whereas the latter is dependent on words and sentences. On this view, sense is prelinguistic meaning that underlies and makes possible signification, which then uses explicit signs as indicative of experiences past, present, and future. So, on this reading, sense is meaningful prior to, and somewhat independent of, language.

Either of these interpretations can be made to work, and they are both plausible accounts of what Dewey claims, in various places, about the relation of language and meaning. The key point, though, is this: whatever one thinks about the relation of language to meaning, it is clear that *sense is a form of meaning*—embodied, experiential meaning. We have defined sense as a prereflective, mostly prelinguistic experience of meaning that situates us meaningfully in our present situation and also makes symbolic understanding of that situation (i.e., signification) possible. The important implication is that meaning is embodied, embedded, and enactive, and not merely an abstract intellectual conceptual or propositional structure lodged in some inner realm called mind. There is meaningful experience that is not directly mediated by language, in the narrow sense of that term that focuses on words and sentences as signs. We have been inclined to favor the option that stresses *sense*, insofar as it draws attention to the many forms of human activity and cultural practice that are profound sources of meaning, without relying, in the first instance, on linguistic signs.

Concepts as Probabilities of Affordances

The dualism that has plagued philosophy for centuries puts mind and body in separate and distinct ontological realms. Concepts, as abstract entities, consisting of sets of features that define an object or event, are regarded as mental, while percepts, arising from bodily engagement with the environment, are physical and bodily. Even those who reject such a strong mind–body form of dualism often retain the dualism that distinguishes different, separate, and independent brain regions for processing perception than for processing conceptualization and reasoning. On this view, percepts are registered as activations in sensory and motor parts of the brain, while concepts are supposedly processed in higher-order association areas of the brain, where abstract representations are created.

Both of these dualisms are mistaken and dangerous. From a naturalistic perspective, there must be a continuity between percepts and concepts, both of which are rooted in sensory, motor, and affective processes in the brain and body. It is best to stop thinking of concepts as any sort of entity whatever, and rather to regard concepts as probability distributions concerning affordances for experience offered by certain objects and events. To illustrate this idea, let's return to the water example. Water, as we saw, affords all sorts of sensory experiences (visual, tactile, olfactory, auditory). Qualitatively, it affords the feeling of wetness, fluidity, hot or cold, clear or murky, and sweet or salty. With respect to motor affordances, it can be poured on plants, mixed with foods, drunk, washed in, swum in, squirted at children, sprinkled on babies at baptism, and manipulated into various forms through application of temperature variations. So, to say that someone has a concept *water*, is to say that, based on past experiences with certain things culturally denominated "water," and based on shared cultural systems of meaning, we have developed a probability distribution of affordances for certain experiences. One of the highest probability affordances would be wetness, even though this quality may not be very salient in some forms of water, such as steam and ice. Water typically affords cleansing, but not when the water is murky, full of sludge, or contaminated. Water typically affords squirt-ability, sprinkle-ability, pour-ability, and swim-ability. In this way we build up expectations of the meaningful affordances of the various things we experience as water.

The category or kind denominated *water* is thus arrayed as what Lakoff (1987) called a radial category structure. Prototypical (Rosch 1975) instances of water are central to the category—as affording highly probable anticipated qualities and effects—while less prototypical instances (e.g., ice, steam, air vapor) are, although in the category, lower-probability expectancies than for the prototypical members. Water as the chemical H_2O becomes relevant, not so much when we are considering primarily the qualitative affordances of a liquid substance, but rather when the chemical makeup is of prime relevance, such as when we want to know how a certain liquid stuff, water, will interact chemically with other substances,

Developing a concept is thus strengthening of functional patterns of neural connectivity based on repeated experiences that generate recurring reliable affordances for organisms with a certain bodily makeup interacting with a structured environment (Barsalou 1999; Martin 2007). Based on prior experience, we project those expectancies for what some object or event will afford us by way of perceptual experience or motor interactions. But, occasionally, we learn new affordances, such as when we discover, in chemistry class, that water applied to a certain metal produces sparks and fire! When water enters into new relations, and thereby offers new affordances, we say that the meaning of "water" is changing and expanding, sometimes in a very minor way, and occasionally in a significant way. Water is chemically somewhat simple, and it is so common and important in our daily lives that we have probably discovered the vast majority of its affordances, so we can expect the concept to remain fairly stable over time, and not subject to radical change. However, stability is not fixity, and most concepts are subject to restructuring resulting from changing conditions, and newly discovered relations.

Stability keeps us appropriately connected to our environment and helps us anticipate developing conditions. Indeed, we rely on fairly stable relationships, and on the internal constraints (supplied by our bodies and brains) to make the assessment of affordances a sane affair—real constraints on conceptual structure and hypothesis formation (Peirce 1878), rooted in functional neural assemblies designed to minimize excessive disruption. This all takes place in a world that generates for us expectancies about affordances that we rely on and internalize as a basis for predictive control.

Too often, we regard concepts as rigidly fixed and circumscribed, when they are actually just highly stable. Lakoff (1987), for instance, showed how the concept *mother*, which was earlier thought of as a completely fixed essential structure, is, in fact, malleable and only relatively stable. Most people think they know what a mother is, but that concept turns out to be far more internally complex than one might think. The development of new reproductive technologies, such as in vitro fertilization, created new types of mothers (e.g., donor mothers, surrogate mothers) that simply could not have existed prior to the creation of these medical technologies. So, while *mother* is a radial category, with the prototypical mother being the female who is married to the father, carries the child, gives birth to it, and nurtures it through life, there are nonprototypical instances of the category, such as birth mothers, donor mothers, surrogate mothers, adoptive mothers, and stepmothers, which do not share all of the defining features of the prototypical mother, even though they are part of the expanding category *mother*. The category *mother* is not frozen or fixed. It is relatively stable, but it has changed in important ways over the past few decades, and we have good reason to anticipate further changes in its meaning, in response to scientific developments.

One important implication of this view of concepts is the rejection of the classical notion of an essence. The essence, presumably constituted by a set of necessary and sufficient conditions of a thing, was supposed to define the features that make it uniquely what it is. This is what Lakoff called classical (or objectivist) category structure. There are perhaps a few essences of this sort, found mostly in artificial systems. Most natural concepts are not essentialist. Dewey thus defined an essence as a set of affordances that we associate with an object or event, based on past experiences and on our current interests and values for engaging that object. The essence of a thing is a set of recurring and fairly reliable characteristics that the object or event affords us. Dewey explains: "when an event has meaning, its potential consequences become its integral and funded feature. When the potential consequences are important and repeated, they form the very nature and essence of a thing, its defining, identifying, and distinguished form" (Dewey 1925/1981, 143).

Concepts are just useful tools for discerning meaning and guiding inquiry, and essences are high-expectancy probabilities associated with certain kinds of things (Gigerenzer 2000, 2007; Todd & Gigerenzer 2007). The problems arise only when we try to transmute these useful conceptual tools—essences—into alleged absolute defining forms: "Essence, as has been intimated, is but a pronounced instance of meaning; to be partial, and to assign *a* meaning to a thing as *the* meaning is but to evince human subjection to bias" (Dewey 1925/1981, 144). Taking recurring, reliable features as absolute fixed essences leads to denial of change within experience, and to the denial of new emergent meanings.

The best way to use the term "essence," then, if we think we must use that term, is as a reminder of characteristics of a thing or event that have proved useful for engaging that object or event. If these categories become rigid and separated out from the ongoing flow of experience, they lose their usefulness as guides to inquiry into nature. Essentialism then becomes, not just an epistemic error, but, more significantly, a moral error, such as when we tie an individual's identity to a cultural stereotype.

Learning the meaning of any particular thing or event is assessing its consequences for experience. Learning new meaning is not merely adding new conceptual content to a preformed and fixed mind. It is a transformation of self, resulting from changing one's brain's structure, through strengthening neural connections, eliminating some connections, inhibiting others, and establishing new ones.

Conceptualization is not, as such, a merely "mental" activity. It is a whole-body activity of ongoing engagement with one's surroundings, dependent on brain and body, and shaped by one's individual and cultural values. Concepts are useful, just as long as they help us orient ourselves in our world. When existential conditions change, so too must our concepts, if they are to remain relevant and useful. This is not a failure of conceptualization, but rather the process by which our concepts can remain meaningful and viable.

Mind Roots in the World

Traditional objectivist theory of knowledge wants to know how mind can ever know a mind-independent physical reality. How could we possibly bridge the gap between what is mental ("inside" mind) and what is physical ("out there" in the world)? The answer, says Dewey, is simple: There is no gap in the first place. Since mind emerges only in and through an organism's ongoing transaction with its environment, mind is always *in* and *of* nature, and knowledge becomes possible through those transactions.

> The world is subject-matter for knowledge, because mind has developed *in* that world; a body-mind whose structures have developed according to the structure of the world in which it exists, will naturally find some of its structure to be concordant and congenial with nature, and some phases of nature with itself. The latter are beautiful and fit, and others ugly and unfit. Since mind cannot evolve except where there is an organized process in which the fulfillments of the past are conserved and employed, it is not surprising that mind when it evolves should be mindful of the past and future, and that it should use the structures which are biological adaptations of organism and environment as its own and its only organs. (Dewey 1925/1981, 211)

This lovely naturalistic passage makes two important points about mind as a natural process. The first is that mind that develops in ongoing engagement with its surroundings will naturally develop structures and habits that pick up on the affordances of its environment. Mind is not a separate disembodied faculty that has to be parachuted into an alien nature, where it stumbles around blindly. We don't have a pregiven and fixed structure of mind that we then have to adapt to the foreign conditions of what is not mind. Instead, mind is the process of nature's development under certain evolutionary pressures and the circumstances of our individual development. Consequently, how we understand, think, and act is a result of the way we have developed through engaging our surroundings.

The second important point is that our so-called mental acts are based on recruitment of processes of sensory and motor experience. Dewey continues, "the mystery that mind should use a body, or that a body should have a mind, is like the mystery that a man cultivating a plant should use the soil; or that the soil which grows plants at all should grow those adapted to its own physico-chemical properties and relations" (1925/1981, 211–212).

These two points are worth dwelling on. The continuity of a mind with a human body, engaging a physical and social environment is what makes it possible to understand one's situation. We are never radically alienated from nature, because we are formed up in and through natural processes—through ongoing interaction with our environment. That is a key part of what makes it possible to "be at home in the world." As we will see later, the traditional problems of how knowledge is possible are based on a mistaken ontological dualism between mind and the natural world. Give up that dualism, and embodied mind can be seen to develop naturally and to learn the meaning of what is experienced.

The continuity of mind and body gives rise to the continuity of "higher" cognitive operations with the supposedly "lower" bodily functions of perception and action. What we call mind recruits operations involved in sensory and motor activities for conceptualization and reasoning.

> Just as when men talk they must use sounds and gestures antecedent to speech, and as when they begin to hunt animals, catch fish or make baskets, they must employ materials and processes that exist antecedently to these operations, so when men begin to observe and think they must use the nervous system and other organic structures which existed independently and antecedently. That the use reshapes the prior materials so as to adapt them more efficiently and freely to the uses to which they are put, is not a problem to be solved; it is an expression of the common fact that anything changes according to the interacting field it enters. (Dewey 1925/1981, 217)

Exaptation of antecedently existing functions to carry out higher cognitive functions is the ecologically parsimonious way of instituting new meanings and experiences. Structures and processes that evolved first for perception and bodily movement are later recruited to shape abstract conceptualization and patterns of inference (Bergen 2012; Berthoz 2002; Buzsaki 2006; Buzsaki et al. 2014; Feldman 2006; Lakoff & Johnson 1999).

We will provide further evidence for this appropriation of perceptual and motor processes for conceptual thinking, but for now the point is that our perceptual concepts involve neural activations underlying the sensory processes involved in having those perceptions, and our action concepts involve activation of neural assemblies for motor and premotor processes involved in the performance of those actions. Instead of evolving distinct brain regions for perception and conception, our brains recruit structures responsible for perception and action to generate concepts and inferential processes.

Habit and the Plasticity of Mind

From a naturalistic perspective, mind and body are not two different things, but rather two dimensions of one ongoing process we can call "body-mind." A dualistic theory of mind posits the mental realm as possessed of its own unique structures and processes of thought, whereas the body is a natural object possessing its own bodily capacities for perception and bodily movement. Once we give up the mistaken notion that there are two different kinds of things interacting here, and, instead, see mind as continuous with body, and thereby using bodily capacities as its means for conceptualization and reasoning, we then have to look to "bodily" processes to explain "mental" ones. In other words, mental processes *just are* bodily processes. Moreover, since the body is a conglomeration of habits of perception, feeling, and action, so, too, the mind is an interpenetration of habits of meaning-making, thought, valuing, and judgment.

A habit is a learned pattern of perception, action, or thought in response to environmental conditions. Habits convert blind energetic impulses into recurrent meaningful patterns of behavior. Since habits arise as patterns of organism–environment interactions, they are both internal dispositions and also incorporate aspects of the environment: "Habits are like functions in many respects, and especially in requiring the cooperation of organism and environment. Breathing is an affair of the air as truly as of the lungs . . . the natural operations like breathing and digesting, acquired ones like speech and honesty, are functions of the surroundings as truly as of a person. They are things done *by* the environment by means of organic structures or acquired dispositions" (Dewey 1922/1988, 15). Once again, we see that habits, like any other allegedly mental dispositions, are actually both in the organism and its environment at the same time, since they are recurring patterns of transaction.

Intellectual habits—habits of thought—are thus in the body and its environment, just as much as they are "in mind." Habits get developed and strengthened through repeated patterns of response to regular structures in the organism's environment. Learning is habit formation. Neurally, habit formation can be modeled as the changing strengths of patterns of neural activation within neural networks.

A critical problem for human learning arises from the role of habit. If learning is habit formation, and if repeated exposure to regular structures in the environment continually strengthens the patterns of neural connectivity in the brain, then how is it possible to learn anything new? It would seem that, in order to acquire new habits, we have to "break" old ones, but those old habits are precisely what have been repeatedly strengthened through recurring patterns of experience. In neuroscience, this is known as the *plasticity dilemma*. Dewey has an answer, of sorts, first recognizing the problem, and then offering his solution: "The very operation of learning sets a limit to itself, and makes subsequent learning more difficult. But this holds only of *a* habit, a habit in isolation, a non-communicating habit. Communication not only increases the number and variety of habits, but tends to link them subtly together, and eventually to subject habit-forming in a particular case to the habit of recognizing that new modes of association will exact a new use of it. Thus habit is formed in view of possible future changes and does not harden so readily" (Dewey 1925/1981, 214).

When new conditions arise, which were not in play in the formation of prior habits, an opportunity is created for release of new impulses that can be taken up into new habits of response. To avoid rigidification of habit, then, one needs to cultivate *more*, rather than fewer, habits. In this way, a given situation will engage a broader range of possible habits, counteracting the tendency to settle back into a single fixed prior habit. One way to understand this process from a neural perspective is to note that individual neurons and clusters of neurons can be involved in multiple functional neural clusters. A given stimulus can therefore activate a number of possible functional clusters, and not just one. This opens up the possibility of new impulses driving the reconstruction of old habits, or the formation of new habits. Dewey describes this whole pluralistic process as follows: "Each habit demands appropriate conditions for its exercise and when habits are numerous and complex, as with the human organism, to find these conditions involves search and experimentation; the organism is compelled to make variations, and exposed to error and disappointment. By a seeming paradox, increased power of forming habits means increased susceptibility, sensitiveness, responsiveness. Thus even if we think of habits as so many grooves, the power to acquire many and varied grooves denotes high sensitivity, explosiveness" (Dewey 1925/1981, 214).

What Dewey is describing here is the process of allostasis. Instead of merely returning to a preset state or value, the organism, in anticipation of changed and changing conditions, develops new habits in a search for a new dynamic equilibrium. The indeterminacy and precariousness of new situations is actually an occasion for recalibrating our expectations and values. It opens the possibility for creatively adapting to novel conditions. This is accomplished, not by miraculous creation of novel ideas, but via bodily processes of allostatic regulation and growth of meaning.

Mind is both a conservation of prior experiences and meanings, but also a continually flowing stream of emergent processes that gives rise to new meaning and possibilities for experience. "For this reason, organic acts are a kind of fore-action of mind; they look as if they were deliberate and consciously intelligent; because, of necessity, intelligent action in utilizing the mechanisms they supply, reproduces their patterns" (Dewey 1925/1981, 215). In this way, mind, as embodied, operates via recruitment of our capacities for perception and action, and, as such, manifests their structure and meaning.

This brings us to the question of how mind, as an ongoing process of shared meaning-making, can involve change. What is consciousness, and how does individual consciousness shape the growth of mind?

5 Consciousness Wakes Up in Mind

We saw how Dewey put human language right at the top of the list of truly amazing wonders of the world. Quite right that he should do so. However, save some room up there for another great wonder, human consciousness! Some folks even think that explaining language is child's play, compared to explaining how conscious experience is possible (Chalmers 1995). That formidable task is even more challenging from a naturalistic perspective that treats consciousness as a totally natural process, thus rejecting any transcendent or supernatural causes, entities, or processes.

People who are critical of naturalistic views of mind ask, "How could consciousness arise out of physical matter?" They hold that no scientific account could ever explain the *experience* of being conscious (Nagel 1979). Owen Flanagan (1991), with a humorous reference to the 1960s one-hit wonder garage band "Question Mark and the Mysterians," has dubbed these folks the "Mysterians." One version of Mysterianism assumes a mind–body dualism, in which mental stuff cannot be explained in terms of physical stuff. But there is a second, more naturalistic version that simply doubts that any neural account of the qualitative experience of consciousness could ever be possible (McGinn 1989). The reason for this, they say, is that no theory of the firing of functional clusters of neurons, or the release of hormones into the bloodstream, could ever account for how it *feels* to be conscious and to experience conscious states, like savoring a taste of key lime pie.

A naturalistic theory of consciousness, like the one we are developing, is typically expected to address at least the following three fundamental questions. (1) How does the mind/brain generate conscious states and give rise to the feeling of conscious experience? (2) Is consciousness merely epiphenomenal, that is, an emergent state of feeling, but one lacking any causal efficacy, or does it somehow also enter into causal sequences in the natural world? (3) If consciousness has causal power, how is that possible, from a scientific perspective?

Since we have to identify the relevant phenomena of consciousness, before we can explain them, any explanation has to begin with some account of what should count as a conscious state (Chalmers 1995). This is no easy task, because different theoretical interests and orientations lead us to circumscribe the relevant phenomena differently, thereby setting us off on different explanatory paths (see Block et al. 1999; Cohen & Dennett 2011; Damasio 1999; Dennett 1991; Flanagan 1992; Gallagher 2018; Koch 2004; Searle 1997, for a survey of phenomena that are often associated with consciousness).

Dewey begins with two commonsense understandings of what consciousness is. The first is simple *awareness* of qualities. The second is understanding the *meaning* of those qualities. On the one hand, the term consciousness "is employed to point out certain qualities in their immediate apparency, qualities of things of sentiency . . . usually termed feelings. . . . On the other hand, consciousness is used to denote meanings actually perceived, *awareness* of objects; being wide-awake, alert, attentive to the significance of events, present, past, future" (Dewey 1925/1981, 226). The distinction here is between merely *having* feelings and grasping the *meaning* of what is felt or qualitatively differentiated. Many animals on the psychophysical plateau are aware of qualitative differences, while animals with mind (rising to the mental plateau) can understand and act on the *meaning* of those qualitative differences. Dewey provides a detailed account of both of these manifestations of consciousness, but, as we will see, his explanations can be supplemented and enriched by recent cognitive science research.

Consciousness as a Transitional Transformative Process

As we noted earlier, the vast majority of our cognitive activity goes on beneath the level of conscious awareness. It all happens "before we know it," so to speak. Our engagement with our environment generates clusters of interpenetrating habits that operate, mostly unconsciously, to shape what we experience, think, know, value, and do. Some of those learned habits are fitted nicely to our present environment, and thereby function automatically and without our knowledge, to create fluid activity in the world. However, in the face of changing existential conditions, habits that were once useful can become ill adapted to the new conditions, leading to a sense of disruption and frustration of fluid action. This frustration of prior habits provides an occasion for instituting the need–search–satisfaction process of biological functioning, with the search phase seeking a way to reconfigure habits to adapt to changed conditions. In humans, operating on the plateau of mind, frustration of habits can also generate the need–search–satisfaction pattern in a cognitive search for the meaning of objects and events.

Recall, from chapter 4, that the need–search–satisfaction process, when it operates in the transition from the psychophysical to the plateau of mind, begins with our *sense* of a situation as meaningful and as affording objects and events relevant to our ongoing experience and thought. Dewey describes this as a vast realm of relevant discriminations and affordances: "Apart from language, from imputed and inferred meaning, we continually engage in an immense multitude of immediate organic selections, rejections, welcomings, expulsions, appropriations, withdrawals, shrinkings, expansions, elations and dejections, attacks, wardings off, of the most minute, vibratingly delicate nature. We are not aware of the qualities of many or most of these acts; we do not objectively distinguish and identify them. Yet they exist as feeling qualities, and have an enormous directive effect on our behavior" (Dewey 1925/1981, 227).

These many types of affordances constitute what Dewey calls the *fringe* of intricately interwoven differentiations that make up, as immanent meaning, the sense of a situation. This fringe of felt sense surrounds and situates the focal meaning active at a particular moment. The *fringe* constitutes the more immediate horizon of meanings operative, in the present, in reference to a focal object or event. This fringe, in turn, is surrounded by an even broader *field* of relatively more stable meanings that are connected to the fringe, and are available, but not currently activated like those in the fringe. So, every thought invokes a vast horizon of possible meanings spreading out continuously from a focal object or event. The focal meanings are the immediate objects of attention, but they are supported by, and get their relevance from, the fringe and field of horizonal significance. Dewey summarizes this threefold relation: "If we consider the entire *field* from bright focus through the fore-conscious, the '*fringe*,' to what is dim, sub-conscious 'feeling,' the *focus* corresponds to the point of immanent need, of urgency; the 'fringe' corresponds to things that just have been reacted to or that will soon require to be looked after, while the remote outlying *field* corresponds to what does not have to be modified, and which may be dependably counted upon in dealing with imminent need" (Dewey 1925/1981, 235–236; italics added).

The focus–fringe–field pattern of functional organization applies generally to perception, conceptualization, and thinking alike. Perceptual events are continuous with conceptual events, so that the transition from perception to conceptual thought is fluid and connected, even though new qualities or properties may arise as conception emerges. This pattern manifests a *principle of continuity* that lies at the heart of naturalistic theories of mind, thought, and language. There are no discrete, independent atoms of either perception or conception. Everything is set within a background of interrelations of objects, qualities, and events. Any given concept or thought involves a field of stable cultural background meanings, a fringe of currently operative meanings of contextual affordances, and an object or event of current focal interest and sustained attention.

This field–fringe–focus pattern also shapes conscious experience. There is a continuity between the field of relatively stable horizonal meanings that operate beneath conscious awareness, the fringe of working meanings of which we are only marginally conscious, and the fully conscious focal object or meaning that is of relevance to our present experience, interests, and values. It is for this reason that William James used the STREAM OF THOUGHT metaphor to capture the fluid, connected, and interpenetrative process of nonconscious processes emerging into conscious experience and thought (James 1890, chapter 9).

This continuous pattern of emergence suggests that consciousness is not typically an all-or-nothing, on–off phenomenon, but instead is a graded affair. Something of which we were previously completely unaware can rise into conscious focal attention, and then, as conditions and interests in the organism–environment interaction change, can pass back out of conscious awareness, sinking back down into the subconscious fringe and field of meaning. For example, in your daily walk, you are listening to one of your favorite tunes, relishing the intricate rhythmic patterns and the lovely melody, when you suddenly become acutely aware of a pain in the instep of your right foot. The previous focus on melody fades from conscious attention and the pain in your foot captures all of your attentional resources. You stretch your foot. As muscle groups release, the pain begins to fade into the background, and you find yourself focusing, once again, on the final sweet notes of the Beatles' "Here Comes the Sun," but perhaps with a slight lingering awareness of an uncomfortable tightness remaining in your foot.

The phenomenology of consciousness thus reveals a general pattern of transition from a field of unconscious habits to fringe awareness of the sense of a situation, to an explicitly conscious grasp of meaning of focal objects and events. The field–fringe–focus pattern is the continuous movement from the nonconscious field of possible relevant meanings, through the unconscious, or marginally conscious, sense of a situation, to the fully conscious awareness of the meaning of a focal object or event. Consciousness is what happens as the meanings of changing experiences are being realized. If mind is the ongoing process of culturally shared meanings developing over time in response to changing conditions, then consciousness is a constructive intervention in the process of meanings that are growing and changing. Consciousness is what makes it possible for an individual person to contribute to the development of mind:

> Mind denotes the whole system of meanings as they are embodied in the workings of organic life; consciousness in a being with language denotes awareness of perception of meanings; it is the perception of actual events, whether past, contemporary or future, *in* their meanings, the having of actual ideas. The greater part of mind is only implicit in any conscious acts or state; the field of mind—of operative meanings—is enormously wider than that of consciousness. Mind is contextual and persistent; consciousness is focal and transitive. Mind is, so to speak, structural, substantial; a constant background and foreground; perceptive consciousness is process, a series of heres and nows. (Dewey 1925/1981, 230)

Consciousness is neither a thing nor a causal force. In its fullest manifestation, it is a process of change in meanings, arising from our ability to feel changes in our body-state in response to changing conditions in our surroundings. That process of realizing new meanings is the second of the two commonplace notions of consciousness discussed earlier. This is possible only for creatures embedded in, and partaking of, mind. The first commonplace notion—consciousness as awareness of qualities—can occur in animals on the psychophysical plateau, because it requires only feeling, and not any system of meanings. Dewey's idea is that some nonhuman animals are conscious in the sense of feeling of qualities, whereas only humans are conscious in the second sense of being aware of the meaning of things and events. Today, as a result of animal research, we might be more inclined than Dewey was to grant certain species of animals the consciousness of meanings, too, and not just of feeling states. That is an empirical matter concerning how many of the processes and capacities in humans are also found, to some degree, in other types of animals (Cabanac 1979).

Consciousness is thus the way an individual person becomes aware of feelings of qualities and their meaning. Dewey nicely illustrates the conscious process with the analogy of what goes on when we read a book or follow the plot of a dramatic play. As we engage the developing narrative, we attend to focal present events unfolding as we read or watch. The meaning of those narrative events depends on the background of the developing narrative, especially in terms of events that transpired earlier, setting expectations about what is happening now, and what is likely to happen in the future, as the play works through to its *denouement*. "When we read a book, we are immediately conscious of meanings that present themselves, and vanish. These meanings existentially occurring are *ideas*. But we are capable of getting ideas from what is read because of an organized system of meanings of which we are not at any one time completely aware" (Dewey 1925/1981, 231). This background *sense* of a situation is mostly nonconscious; however, this sense is what eventually makes possible the consciousness of ideas involved in focal signification—the use of signs to evoke and enact meanings. Dewey concludes: "The larger system of meaning suffuses, interpenetrates, colors what is now and here uppermost; it gives them sense, feeling, as distinct from signification" (Dewey 1925/1981, 231).

We tend to think of consciousness as a moment of intense focal awareness—the bright light of attention shining to illuminate a focal experience or meaning. But consciousness fades off into a marginally or wholly nonconscious sense of the immanent meaning of a situation (i.e., its sense). The fact that sense provides a vague, holistic, and affect-rich contextual grasp of a situation does not make it any less real, or any less crucial for meaning, than are the more explicit, fully conscious, selected objects of focal attention. Dewey often observed, especially in his book *Logic: The Theory of Inquiry* (1938/1991), that neglect of the vague, in favor of the precise and explicit, is one of the great mistakes of philosophies that focus only on the cognitive dimensions of thought, at the expense of the aesthetic qualitative dimensions. The felt sense of the vagueness of a situation is just as real, and just as important, as the highly differentiated and attended to aspects of an object or event.

The phenomenologist philosopher and psychotherapist Eugene Gendlin (1991) developed a "philosophy of the implicit," focusing on the felt sense that underlies and shapes all our implicit, reflective, conceptual, and representational processes of thought. He provided ways to engage the prereflective dimensions of thought that situate our thinking in the world and constitute both what and how we think, even as we may be unaware of those conditions. He often illustrated this preconceptual enactment of meaning with the example of a poet searching for a line to end a quatrain. She tries out now this and now that concluding line, but each line somehow fails to capture the *that* (the blank one is trying to fill) that would better carry forward the meaning that is developing in the poem. When she hits upon a line that feels right, and that realizes the meaning that is working in the quatrain (and the whole poem as it develops), there is a felt release and a sense that a thought is being brought to consummation. Gendlin cautioned us not to think of the felt sense as a pregiven meaning that somehow has to be represented by a conceptual or propositional structure. Rather, the prereflective and the conceptual/representational are intimately intertwined in one growing process of meaning-making. Or, as Dewey might put it, *sense* and *signification* are intertwined, interdependent, and co-developing.

What we are describing here is the ongoing, ever-developing process of meaning-making that moves from the nonconscious or dimly conscious sense of a situation to the mostly conscious awareness of meanings at work in transforming a situation through signification. Past events and meaning are being carried forward into the present and then projected as expectancies for future experience:

> Thus the purport of past affairs is present in the momentary cross-sectional idea of a way which is more intimate, direct and pervasive than the way of recall. It is positively and integrally carried in and by the incidents now happening; these incidents are, in the degree of genuine dramatic quality, fulfillment of the meanings constituted by past events; they also give this system of meanings an unexpected turn, and constitute a suspended and still indeterminate meaning, which induces alertness, expectancy. It is this double relationship of continuation, promotion, carrying forward, and of arrest, deviation, a need of supplementation, which defines the focalization of meanings which is consciousness, awareness, perception. (Dewey 1925/1981, 232)

Consciousness, according to this account, is the feeling of changes in one's body-state as it engages its environment. It is our feeling of being affected by objects of perception and thought. Dewey succinctly summarizes his definition as follows: "Consciousness, an idea, is that phase of a system of meanings which at a given time is undergoing re-direction, transitive transformation. . . . Consciousness *is* the meaning of events in course of remaking" (Dewey 1925/1981, 233).

The Mysterians fancy themselves the champions of consciousness because they see it as either the feeling of an irreducible quality of some experience, or else as a causal force entering into and affecting the transformation of experience. The Mysterians thus claim that science cannot explain why felt changes feel the way they do, and some of them regard consciousness as an originating causal force capable of affecting the course of experience (Taylor 1963).

Dewey dismisses both of these claims as creating pseudo-problems that set us on paths of inquiry that are ill conceived and misleading. First, our inability to explain how and why felt changes feel the way they do is a matter of the limitations of human language, rather than a profound epistemic problem: "It is impossible to tell what immediate consciousness is—not because there is some mystery in or behind it, but for the same reason that we cannot tell just what sweet or red immediately is: it is something had, not communicated and known" (Dewey 1925/1981, 232). Using the term "red" of an apple, "sharp" of a musical note, or "sour" of a lemon may make it possible to identify a specific quality of some object or event, but it cannot explain how that red, sharp, or sour feels, or is experienced. That our language is incapable of capturing the fullness and richness of our experiences is a fact about the limited semantic resources of natural languages, and not an indication of some profound mystery about the reality and nature of qualities.

The idea of consciousness as a causal force is equally mistaken and misleading. Some Mysterians believe that consciousness somehow *causes* us to have certain experiences. We argued earlier that a cause is not an independent force, but just a characteristic dimension or contributing condition, or set of conditions, of an experience—conditions that we select out for attention, based on our needs, interests, and values. Every "effect" is potentially part of a further set of conditions that jointly bring about a new complex existence. Therefore, events that in one context are regarded as effects can become, in another context, contributing causes of future events. Concerning consciousness, then, "its 'cause' is only the fact that this is one of the ways in which nature goes on. In a proximate sense of causality, namely as place in a series history, its causation is the need and demand for filling out what is indeterminate" (Dewey 1925/1981, 233). Consciousness is simply a qualitative event in which a cluster of conditions are undergoing change. There is no single "cause" of that change, because it is the result of the whole set of relevant conditions operating together. Consciousness, though part of a particular experience, is in no sense "the cause" of anything. We will return to this issue later in this chapter, but, for now, the point is to explain why we should keep consciousness—a real phenomenon—from being recruited, as an inexplicable force, into the realm of the Mysterians.

Consciousness and the Making of the Self

Most folks think of themselves as a unique bodily individual with a unique mind. That mind is private—an inner mental realm in which we have a distinctive perspective on, and ownership of, our thoughts and feelings, which are not immediately accessible to other people with their individual minds. We also see ourselves as agents, possessed of free will, and capable of acting in and on the world.

This received view captures some important aspects of our selfhood, but it misses others. For one, while it recognizes that I have "my" mind, and "my" body, which generate *my* awareness of *my* perceptions, feelings, and thoughts, while you have "your" mind, with *your* personal feelings and thoughts, it overlooks the ways in which each individual self is constituted always in and through its relations with others, in a community of shared meanings, practices, and values. In this respect, mind is a developing system of shared meanings, within which individual persons grow, and from which they get the culturally stabilized meanings that allow them to make some sense of themselves and their world.

Second, the commonsense view misses the fact that the vast majority of our cognition goes on unconsciously (e.g., Feldman 2006; Helmholtz 1873; Johnson & Tucker 2021; Rozin 1976). Therefore, most of what goes into the making of our selfhood is not even available to our introspection, and we have only a very limited conscious control over our identity. Consciousness is an emergent awareness of changes in the body in response to changing existential conditions, but it is certainly not "running the show" most of the time. Our allegedly *free* will turns out not to be nearly as free as we might think or hope, and consciousness is just part of the way we participate in events, both events in the internal milieu and those in the social milieu (Eagleman 2011).

So, we have to ask, "What is the self?" and "What role does consciousness play in the formation of the self?" We will address the nature of the self more fully in chapter 7, but for now we will focus chiefly on consciousness, with only an abbreviated investigation of its relation to selfhood. Even though Antonio Damasio makes no explicit reference to Dewey, his account of levels of consciousness and selfhood captures many of Dewey's views. Damasio's account is richer and more nuanced than Dewey's, insofar as it draws on extensive neuroscientific evidence that was not available in Dewey's day.

Damasio defines consciousness as "a state of mind that occurs when we are awake and in which there is private and personal knowledge of our own existence, situated relative to whatever its surround may be at a given moment" (Damasio 2010, 158). When Damasio describes consciousness in terms of a personal and private knowledge, we must not think of knowing as some sort of intellectual operation "in the head." Rather, consciousness is a whole brain–body ongoing engagement with one's surroundings that is integral to our more complex functioning in the world. It plays a role in developing expectancies concerning what we will encounter at the sensory and motor interface with our environment (Seth 2013). At certain higher levels of complex interactions, this process involves ordinary bodily sensibility, regulation of the internal milieu, planning for action, and adjudicating and managing the social milieu.

Many researchers have suggested that, across many animal species, we find a link between consciousness and some sort of sensation (Cabanac 1979; Denton 2005; James 1890; Panksepp 1998). Animals, from invertebrates to vertebrates, have systems for sensing gradients (e.g., of intensity, quality, or availability) in relation to valued objects, setting the context for behavioral response (Maier & Schneirla 1935/1964; Young 1959). These sensory value assessments are fundamental for survival and flourishing. Therefore, bodily sensibility—the body's active evaluative engagement with its surroundings—is not an abstraction (Schulkin 2004). Indeed, bodily sensibility is the precondition for the capacity to assess which options for action aid continued viability. It constitutes the basis of an organism's functioning and well-being. This adaptive process often requires assessment of the value of response options relative to organism viability. This body-based continual evaluative activity occurs via the fluid interaction of both peripheral (e.g., perceptual) and central nervous systems in the body–brain, both of which are permeable and transactive, as they engage conjointly in adapting, coding, responding, and remembering events that matter for the well-being of the organism.

Following a line of many investigators (e.g., Herrick 1929, 1948; James 1890), Damasio (1999) claims that all experience begins with a bounded organism in ongoing interaction with its surroundings. The supreme value for any organism is maintenance, within the boundary of the organism, of the conditions necessary for life functions (Bernard 1865/1957; Cannon 1932/1963; Richter 1943). As part of this process in animals with brains, there are evolved neural architectures for monitoring changes in bodily states in response to the animal's engagement with its surroundings.

Recall that this process involves the brain's making of neural maps of our internal bodily states, of the objects in the external world, and of the ways our body-states are changed as we interact with our surroundings. The brain generates neural maps of both internal states and external objects, places, and events (Swanson 2003a, 2003b), rendering engagement with objects fundamental to bodily regulation (Barrett & Simmons 2015; Seth 2013). These maps preserve the important structural relations of what is being mapped. For instance, in visual (retinotopic) maps, adjacent areas in a visual array, as realized within the retina, are preserved in the relations of neurons, or clusters of neurons, in the brain's visual processing regions (Sterling & Laughlin 2015). In tonotopic maps, adjacent pitches are realized in adjacent neural areas in auditory cortex. Neural maps are simply patterns of neural activation, created by the organism's sensory–motor interfacing with its surroundings. As such, we are not conscious of the firings of the individual neurons in the functional clusters of neurons that constitute the map. However, activation of maps can sometimes give rise to what Damasio calls *images*, such as a visual image of a cat chasing a mouse. Damasio explains that "when I use the term *image*, I always mean *mental* image. . . . When I refer to the neural aspect of the process, I use terms such as *neural pattern* or *map*. Images can be conscious or non-conscious" (Damasio 1999, 317–318). Moreover, images are not just visual, but can be generated by any sensory modality, such as touch, taste, hearing, or proprioception. The *mental image* of a spatial relation, a relation of pitches, or a gradient of textures is the result of the neural maps in the respective sensory, motor, and affective systems of the brain.

Damasio sees consciousness as a flow of images generated by our engagement with objects in our surroundings (both external objects and internal bodily states). The term "image" is somewhat problematic, given the semantic history of its use as an alleged inner perception, representation, or idea that somehow mirrors some external object, event, quality, or relation. Rather, we should think of body, brain, and mind as continuous with one another, and an image is simply the way in which we are aware of changes in our brain and body-state, with this continuous process requiring constant monitoring and maintenance. Images are not mental entities dancing on the inner mind-stage of Cartesian consciousness. Neural maps and their corresponding images are not two different entities or processes that somehow must be combined by the mediation of some third thing or force. The images are just the result of the activation of neural maps and their associated affective valences. Consequently, maps do not separate us from the world we dwell and act in, but rather constitute our ways of having and being in a world. Neural maps are the basis in the brain for mental images (e.g., Kosslyn 1980, 1994), so that "neural map" and "image" indicate two intertwined aspects of one process, which is the experience of a mental image based on activation of a specific neural pattern.

Core Consciousness

Damasio distinguishes two types of consciousness based on the scope of what is attended to. *Core consciousness* is the limited scope of having mental images in a moment of present awareness, such as seeing the delicate, joyful daffodils in the vase on the table before you. *Extended (or autobiographical) consciousness* is broad in scope, encompassing not just the present moment, but prior remembered events and projected future ones. It is the extended moments of experience over time, history, and context that constitute the episodic and continued memories of individuals (Schacter 1996, 2019; see also James 1890).

Damasio summarizes these two modes of consciousness as follows: "The minimal-scope kind I call *core* consciousness, the sense of the here and now, unencumbered by much past and by little or no future. It revolves around a core self and is about personhood but not necessarily identity. The big-scope kind I call *extended* or *autobiographical* consciousness, given that it manifests itself most powerfully when a substantial part of one's life comes into play and both the lived past and the anticipated future dominate the proceedings. It is both about personhood and identity. It is presided over by an autobiographical self" (Damasio 2010, 168–169).

The distinction between core and extended consciousness is not meant to be a radical ontological difference, but rather to indicate a temporal continuum in which a moment of present awareness (core consciousness) is situated in relation to our memory of past experiences and anticipation of future potential experiences (extended consciousness) (Donald 1991; Schacter 1996). Core consciousness corresponds loosely with Dewey's account of a momentary feeling of a change in our bodily state, while extended consciousness corresponds to use of shared meaning resources in remembering past events and anticipating future events. Damasio speculates that many animals have core consciousness, since it requires only a capacity for feeling changes in their body-state, while probably only a very few species are likely to have extended consciousness, because it requires an ability to understand the meaning of events (Damasio 2010, 171–172; see also Denton 2005; Panksepp 1998).

How are these two types of emergent consciousness possible? Damasio proposes that it is a result of the brain's ability to make neural maps that are experienced by us as mental images (Kosslyn 1980, 1994). As a result of our brain's map-making, we have images of qualities and relations in and among external objects, but we also have images of our bodily states and the changes in those states.

Consciousness is a product of images realized by neural maps. Damasio thus suggests that "conscious minds arise from establishing a *relationship* between the *organism* and an *object-to-be-known*" (Damasio 2010, 188). Core consciousness is what allows us to know what we are experiencing. It generates "the feeling of what happens" (Damasio 1999) and is the key to the process by which "self comes to mind" (Damasio 2010). The mechanism for this has three components; (1) we make *first-order maps* of external objects and locations, (2) we make *first-order maps* of our current bodily states, and (3) we make *second-order maps* of how our body-state is being changed by our engagement with objects in its environment. It is this second-order awareness that constitutes core consciousness. Damasio summarizes this emergent process as follows: *"Core consciousness occurs when the brain's representation devices generate an imaged, nonverbal account of how the organism's own state is affected by the organism's processing of an object, and when this process enhances the image of the causative object, thus placing it saliently in a spatial and temporal context*" (Damasio 1999, 169; italics in original).

This feeling of what is happening is constituted by a flow of images generated by our ongoing temporal engagement with various objects, either physical objects or memory objects of past experiences. Every time we experience a different object, the body-brain undergoes change, as the brain readjusts the body to its new encounter with the object. When this happens, the object experienced gains temporary salience in attention, and we have a feeling of knowing the object, just insofar as we feel, via a second-order mapping, what is happening at the moment. In addition to *feeling* these changes in our body-state, we come to experience them as in us, as *ours*, which gives us our sense of "ownership" of the images we have in mind.

A crucial consequence of this account is that consciousness occurs just so long as we have a series of images of experienced objects and events that gain attentional salience against a background horizon of possible experiences. When we cease to feel those changes in our body in response to imaged objects, we lose consciousness, and fall back into the preconscious workings of the brain as it seeks to maintain organism homeostasis and allostasis, to continue to remain viable.

Damasio summarizes the emergence of core consciousness: "What is being added to the plain mind process and is thus producing a conscious mind is a series of images, namely, an *image* of the organism . . . ; the *image* of an object-related emotional response (that is, a feeling); and an *image* of the momentarily enhanced causative object. *The self comes to mind in the form of images, relentlessly telling a story of such engagements*" (Damasio 2010, 203). Around 130 years ago, William James (1890) explained consciousness as a series of continuous pulses generated by experienced objects. He described a moment of present conscious experience as "the specious present" which emerges from a prior flow of experienced objects and then projects toward future experiences. He thought of this momentary unit of consciousness as spreading over a roughly twelve-second time period, thus determining how much a person can consciously entertain in one pulse of consciousness. Whether that is an accurate account of the time interval is an issue for empirical study, but the point about pulses of consciousness is a major insight. There is no inner homunculus that entertains mental images prancing in some inner mental realm. There is no need to summon a homunculus to be the locus of consciousness. Instead, consciousness is sustained by the body's monitoring of the changing of its body-states in response to objects encountered in experience. We experience the feeling of what is happening just so long as there are objects encountered that generate body-state changes. Consciousness comes in a series of pulses. Its relative unity and stability are just the consequence of the fact that these pulses are in and of an embodied self that is going about the business of preserving the homeostasis and allostasis necessary for survival and well-being, thereby preserving a relatively stable internal milieu.

Extended Consciousness and the Autobiographical Self

A pulse of core consciousness is a momentary and fleeting event. If a person is capable only of a moment of consciousness, but lacks any memory of past events, we suspect some pathology in the brain. One of the wonders of the human brain is our rich and extensive capacity for memory that permits us a meaningful narrative of the course of the self over time. Instead of a knife-edge present moment of core consciousness, we also are able to reach back into times past and to carry them forward, both as remembered here and now, and also as a meaningful basis for future experiences and actions. The result is *extended consciousness* and an always developing narrative of the self as experiencer and actor in the world.

Franz Brentano (1874/1973) observed that all mental acts have a content—an object toward which they are directed. He called this relation *intentionality*, following the use of that term in medieval philosophy. Intentionality is the "aboutness" of a conscious mental state. Damasio argues that in core consciousness the intentional object is typically some physical thing or event that we experience through our senses. However, extended consciousness also has as its objects memories of past experiences. Consequently, just as a moment of core consciousness requires a continuous succession of experienced objects, so also extended consciousness requires a continuous flow of memory objects or images of past events. The resulting *autobiographical self* is, although somewhat stable, never a fixed or completed object, because it is always being remade, as new experiences arise and memories are reconfigured and acquire changed meanings.

Damasio proposes that "constructing the autobiographical self depends on two conjoined mechanisms. The first is subsidiary to the core self mechanism and guarantees that each biographical set of memories is treated as an object and made conscious in a core self pulse. The second accomplishes a brain-wide operation of coordination that includes the following steps: (1) certain contents are evoked from memory and displayed as images; (2) the images are allowed to interact in an orderly manner with another system elsewhere in the brain . . . ; and (3) the results of the interaction are held coherently during a certain window of time" (Damasio 2010, 212–213).

In other words, just as physical objects can affect a change in our body-state, thereby altering the self via new meanings, so also can memory images reconfigure our self-identity, as we become conscious of our changing body-states. In this case, the "objects" that sustain consciousness and achieve enhanced salience are memories that form our narrative sense of self. We experience ourselves as protagonists co-creating the narrative arc of our temporal existence.

Damasio provides a quite detailed theory of the neural basis in the brain for extended consciousness (see Damasio 2010, 212–240). The key to his theory is the notion of a supramodal "Convergence–Divergence Region" (a CDRegion), in which maps of first-order sensory, motor, and affective functions can be combined to generate second-order maps of complex experiences. For example, the associating of sensory images from multiple sense modalities (e.g., vision, touch, hearing, taste) occurs in a convergence region that combines these diverse sensory images into a unified stable multisensory image (Gregory 1966; Gross 1999; Hubel & Wiesel 2005; Rolls 1996, 2000; Rolls & Treves 1998).

In the case of creating a unified and coherent life narrative, it is memory objects (i.e., memories) that are associated in second-order Convergence–Divergence regions thereby building an autobiographical self-narrative. In the ongoing construction of our autobiographical self, "these CDRegions hold records of previously acquired knowledge regarding the most diverse themes. The activation of any of these regions promotes the reconstruction, by means of divergence and retroactivation into image-making areas, of varied aspects of past knowledge, including those that pertain to one's biography" (Damasio 2010, 216).

The autobiographical self is thus not a fixed entity of any sort. It is a relative stability in a flow of changing meanings about the experiences of the self. It is a self-in-process that continually changes and reconstitutes itself. An important goal of self-maintenance is to cultivate habits of intelligent appraisal and remaking that sustain this identity. Both explicit and implicit memory mechanisms underlie this capability (Schacter 2019).

A key element in the constitution of a conscious self is the mechanism for selection of which images are allowed to rise to the level of conscious awareness. Most of our cognitive processing is nonconscious and not subject to reflective control (Feldman 2006; Kahneman 2011; Lakoff & Narayanan in press). The brain is continually generating a prodigious number of patterns of neural connectivity, but only a select few of them become candidates for conscious attention. In the organism's quest for viability, those patterns that tend to be most relevant for sustaining homeostasis and allostasis, as necessary for life maintenance, are likely to be marked via feelings and emotions. Damasio summarizes his thesis: "The strategy was to select automatically those images that were most valuable for ongoing life management—precisely the same criterion presiding over the natural selection of the image-making devices. Especially valuable images, given their importance for survival, were 'highlighted' by emotional factors. The brain probably achieves this highlighting by generating an emotional state that accompanies the image in a parallel track. The degree of emotion serves as a 'marker' for the relative importance of the image. This is the mechanism described in 'the somatic marker hypothesis'" (Damasio 2010, 174–175). The somatic marker hypothesis makes use of recognized functional brain regions and processes involved in sustaining the viability of the animal. Moreover, the anatomical revolution in neuroscience demonstrated that neocortical sites had direct neural and chemical connectivity to brain stem sites that regulated core bodily responses, and also direct neural contact with peripheral systems in the alimentary canal (Swanson 2003a, 2003b). There was no separation between bodily regulation and neocortical or cortical probing for adaptative responses (Schulkin 2003, 2004; Tucker & Luu 2012).

This account makes sense of why consciousness should have arisen, and why it prevailed in certain animals. The suggestion is that "generating, orienting, and organizing images of the body and of the outside world in terms of the organism's needs, increased the likelihood of efficient life management and consequently improved the chances of survival. Eventually, consciousness added the possibility of *knowing* about the organism's existence and about its struggles to stay alive" (Damasio 2010, 175). No doubt, Charles Darwin is blushing with endless delight. Indeed, within a short time after the great book *On the Origin of Species* appeared in 1859, the search was on to explore the biological and neural bases of sensing the world, the diverse sorts of emotions that aid survival, social contact, and avoidance behaviors, and the levels of the nervous system where consciousness is realizable (see Jackson 1884/1958; Sherrington 1906/1948; Von Holtz 1973). We now understand that the neural circuits that underlie detection of threat or danger are obviously critical and reach into all aspects of bodily sensibility, from peripheral detectors to more central processing systems, like the amygdala (LeDoux 1996, 2015; Rosen & Schulkin 1998). For Darwin, there was no "Descartes' Error" (Damasio 1994), no divorce from the body. This intertwining of visceral needs and higher cortical processes was appreciated by a broad range of biologically oriented investigators (Dewey 1896; Hebb 1949; James 1890; Richter 1943; Stellar 1954).

How Is Consciousness Possible?

The experience of being conscious is one of the most visceral, intimate, and self-constituting experiences we have. Without it, we would languish in a state with little or no functional awareness, persisting via unconscious homeostatic and allostatic processes. We wouldn't know what was happening to us, who we are, or what we are doing. Yet, despite the profound importance of consciousness for our lives, we cannot seem to get a really phenomenologically satisfying and scientifically respectable theory of its nature. Indeed, for several decades in the twentieth century, many philosophers, psychologists, and neuroscientists instinctively shied away from the subject of consciousness, often claiming that it was a purely subjective phenomenon for which no objective scientific account could be adequate. Declaring that conscious experience was not a fit topic for behaviorist scientific research, they opened the door for the Mysterians, who hold that there can be no scientific explanation of what it feels like to have a conscious state.

Over the past couple of decades, however, rapid advances in cognitive neuroscience have made the subject of consciousness respectable and even popular once again, resulting in the attitude that one cannot have an adequate theory of mind without a correlative theory of the nature and neural basis of consciousness. Having briefly surveyed some of Damasio's chief hypotheses about the nature and possibility (from a neuroscience perspective) of core and extended consciousness, we now want to supplement this with Gerald Edelman's and Giulio Tononi's theory of the neural conditions for consciousness, as one promising model for an entirely naturalistic perspective that emphasizes bodily sensibility and neural functions.

We saw, with Damasio, that consciousness is our feeling of changes in our body-state, as our body interacts with objects and events in our environment. Consciousness is the feeling of how those experienced objects are changing our body-state. The brain makes maps of external objects, maps of the state of our bodies, and maps of how those objects affect our current body–brain states (see chapter 7). Core consciousness is the "narrative" of the drama of our constant processing of our engagement with our world. This narrative consists of spatial and temporal relations providing a "feeling of what happens."

The brain's "maps" are patterns of weighted connections among neurons and clusters of neurons (Sporns 2011). However, there is no inner homunculus that makes these maps, nor is there any central executor that could organize and temporally sequence those maps into a flow of conscious experience (Konorski 1967; Penfield 1959). Instead, the multifarious regions of the brain–body that are responsible for creating these diverse kinds of neural maps tune their maps through interconnections and communications across various functional brain regions. Unlike a digital computer, which has a central clock and sequentially orders meaningless symbols, human brains operate through massive parallel processing, in which different brain regions reciprocally influence one another, in an ongoing process of revisions of neural connectivity and connection strengths (Herbert & Schulkin 2002; Norgren 1995; Saper & Loewy 1980; Saper et al. 2005; Sporns 2011; Swanson 2003b).

Edelman calls this type of processing "reentry" (Edelman 2004; Edelman & Tononi 2000), and others have used the apt name "reafferent processing" (Gallistel 1982; Sperry 1961; Von Holst 1973; Weiss 1939). Through this ongoing process of reentry, many functional brain regions act together to create changes in their various types of neural maps, with the mutual influence, coordination, and communication occurring without any central associative center or faculty to control the tuning up of functional neural clusters (Weiss 1939): "During development, large numbers of reciprocal connections are established both locally and over long distances. This provides a basis for signaling between mapped areas across such reciprocal fibers. Reentry is the ongoing recursive interchange of parallel signals among brain areas, which serves to coordinate the activities of different brain areas in space and time. Unlike feedback, reentry is not a sequential transmission of an error signal in a simple loop. Instead, it simultaneously involves many parallel reciprocal paths and has no prescribed error function" (Edelman 2004, 39–41).

Edelman's hypothesis is that it is the brain's ability to create various kinds of neural maps, coupled with its capacity for massive reentrant processing of those maps, that gives rise to a momentary conscious experience (which he calls a moment of primary consciousness). He speculates that most of this reentrant processing occurs within regions of the thalamus as it projects into and interacts with various cortical areas. For many years now, it has been helpful to conceive of the thalamus as a set of pathways for channeling various types of sensory input to diverse functional regions of the brain's cortex (Rolls 1996; Saper et al. 2005). However, we have now discovered diverse ways the brain facilitates cortical activity outside of thalamic–cortical connectivity alone—other subcortical pathways for these functional clusters besides the thalamus, especially via particular chemical signaling systems that project to cortical regions for most sensory information (Herrick 1905, 1948; Norgren 1995).

Edelman names this system of ongoing thalamic–cortical reciprocal processes the *dynamic core*, which is:

> the reentrant structure capable of integrating or binding the activities of the various thalamic nuclei and the functionally segregated cortical regions to produce a unified scene. . . . Changes of state in the dynamic core in response to signals from within and without engage new sets of dynamic functionally segregated circuits in short times, and this property accounts for the differentiation of successive scenes constituting the conscious state. . . . Qualia *are* these discriminations. The great variety of discriminations emerges because the dynamic core is a complex system that can maintain functionally segregated parts while integrating the activity of these parts over time in a very rich combinatorial fashion. (Edelman 2004, 70–72)

There is a great deal that needs unpacking here but let us consider some key implications of the dynamic core hypothesis. First, "the dynamic core, with its millisecond-to-millisecond utilization of an extraordinary complex of neural circuits, is precisely the kind of complex neural organization necessary for the unitary yet differentiable properties of the conscious process" (Edelman 2004, 69). This multidimensional reciprocal activation of reentrant neural processes generates the discriminations that *are* conscious states.

Second, it is widely recognized that conscious experience depends on at least two major conditions. (1) It occurs only when there are multiple diverse functional brain areas operating together, through reentry, thus providing *differentiation* of functional processes and perceptual content. (2) These diverse functions are *integrated* within a temporal window, to constitute a unitary moment of core consciousness. So, core consciousness requires both discrimination of differences, but also the temporal integration of those differentiations into a unified moment of time.

Regarding the first condition, the existence of massive ongoing reentry creates exactly the differentiations among various sensory processes (e.g., visual, auditory, tactile, somatosensory, proprioceptive, and kinesthetic) and motor processes (e.g., walking, grasping, throwing, lifting, reaching) that allows us to discriminate the feeling of a particular quality. Edelman notes that, in order to perceive a single quality (e.g., the redness of a rose, the smell of a rose, the sound of a church bell, the taste of a lemon) we need the ability to discriminate that quality from a host of different qualities. The redness of the rose is felt differently by contrast with the taste of a lemon, or the smell of fresh-mowed grass. He concludes, "But how do we know that the color qualia are in fact what they are, except in the higher-dimensional space that also maps the various other qualia, thus allowing these mutual distinctions" (Edelman 2004, 65). Moreover, since the various types of sensory experience route at least partially through different brain architectures, it is not surprising that they are felt as qualitatively different, as in the difference between the taste of dark chocolate, seeing the red head of a hummingbird, or smelling the aroma of a pinot noir.

The second characteristic condition for consciousness—integration within a narrow temporal window—is also explained by the "communication" among multiple functional clusters of neurons. The complex mutual tuning of maps across multiple regions constitutes the only stability there is for a moment of time. For, as we constantly encounter new conditions and our situation changes, we have to reconstitute a relative dynamic equilibrium that is itself fleeting, only to be replaced, in the next moment, by the next integration of differentiated qualities.

A third, absolutely crucial, implication is that "a dynamic core is therefore a process, not a thing or a place; and it is defined in terms of neural interactions, rather than in terms of specific neural locations, connectivity, or activity. Although the dynamic core will have a spatial extension, it is, in general, spatially distributed, as well as changing in composition, and thus cannot be localized to a single place in the brain" (Edelman & Tononi 2000, 144;

see also Koch 2004; Tononi & Koch 2008; Wolf 1984). The dynamic core that supports consciousness is a very temporary organization of functions. It is *bodily*, but it cannot be reduced to any unique set of neural connections. A correlate of this fact, as we will see in chapter 7, is that the self is never a fixed or completed object, but rather a pattern of dynamic stabilities in the ongoing flow of experience. This latter characteristic is the result of what is known as the "degenerate" nature of neural systems, by which is meant that the brain generates more than one set of patterns of neural connectivity that can give rise to the "same" quality of a conscious state.

A fourth major corollary of the dynamic core hypothesis is that the generation and maintenance of the patterns of reentrant connection that give rise to the dynamic equilibrium is shaped, at all times, by fundamental values of the living creature. These values are determined by the makeup of the organism (its body and brain), the organism's basic needs, the affordances of its environment, and its developmental history. Evolution provides brain architectures that generally serve the animal's survival and viability. The path of individual development supplies additional values and interests, and therefore shapes what we attend to and how we respond to changing conditions.

What Is Consciousness Good for, Anyway?

These conditions and characteristics of consciousness suggest that it serves certain important functions related to our well-being. It enables awareness of our surroundings, of the state of our bodies, of the meaning of our experience, and of the affordances for possible actions relative to our present situation. However, awareness of these important factors does not immediately translate into action, so we are led to wonder whether consciousness is merely an epiphenomenal awareness of what is happening, but without any causal influence on what happens.

Consciousness is the result of the operations of a dynamic core that makes possible discriminations of qualia: "Conscious processes arise from the enormous numbers of reentrant interactions" (Edelman 2004, 77). Because these discriminations (as qualia) arise mostly from sensory inputs shaped by the body's value systems, are tied to motor programs, and are influenced by emotional responses, we tend to experience consciousness as located in a unified self, rooted in a physical body. In core (primary) consciousness "this self exists in a remembered present, reflecting the integration of a scene around a small interval of time present" (Edelman 2004, 77). This reentrant dynamic core of integrated discriminations manifests as what Edelman calls a "phenomenal transform"—the feeling of what is happening in your body–brain as it engages its surroundings.

In order to explore the issue of whether this phenomenal transform has any causal efficacy, we must consider the relation between conscious states and the neural processes underlying them. Edelman's argument can be summarized as follows. Let *C* stand for a conscious state or process, the phenomenal transform. Let *C′* be the underlying neural processes, consisting mostly of the ability to make neural maps, such as first-order maps of things in the external world and of our current bodily states, and second-order maps of how engagement with those external things or events gives rise to changes in our body-states. Then,

- *C is entailed by C′*—Entailment means that where a particular *C′* state exists, we should expect to find a correlative *C* state.
- *C′ is causal*—Neural states and processes causally influence one another.
- *C does not cause C′*—Conscious states and processes do not cause anything. Instead, they exhibit relations with other conscious processes (such as becoming aware that the grumbling in your stomach indicates that you are hungry).
- *C is the only reliable information we have concerning our C′ states*—Because any given conscious state *C* is correlated with some set of neural processes, *C* gives us insight, though typically quite modest and very limited, of what's going on in our bodies and in the world we inhabit. This is not perfect or infallible knowledge, but in general it is a highly reliable indication of what's happening at the *C′* level.

Accordingly,

- "Consciousness C as a property of C′ is a reflection of the capacity to make refined discriminations in a multidimensional qualia space. This phenomenal transform, reflecting events in that space, is a reliable indicator of the underlying causal C′ events. The consequence of this line of reasoning is that evolution selected C′ (underlain by the neural activities of the dynamic core) for the efficacy in planning conferred by its activity. At the same time, however, such C′ activity entailed corresponding C states. Indeed, there is no other way for an individual animal to directly experience the effects of C′." (Edelman 2004, 79–80)

Although Edelman does not explicitly make this point, we would suggest, following Dewey, that what consciousness primarily does for us is to make possible our understanding of the meaning of our experience. Although core (or primary) consciousness alone is not sufficient to supply the meaning of all events, it plays a key role in our ability to discern the implications for experience given by an experienced object or event. As our memory expands and is enriched, we experience extended consciousness of new meanings of events and new expectancies for present and future experience. Consequently, although consciousness does not causally affect neural processes, it can offer us important insight into what is happening to us at any given moment. While it is true that the vast majority of our cognition and experience operates unconsciously, we nevertheless do have the ability to explore the meaning of a particular situation in a way that reveals values and affordances for transformative action. There is no inner homunculus, no executive center of agency, and no transcendent ego running the show and causing action. But there is, via consciousness of meanings (both presently enacted and future possibilities), understanding of the implications of things, experiences, and actions, as a basis for intelligent redirection of experience.

Something like the account just examined seems to be the only way to have a fully naturalized view of mind and consciousness. It recognizes that all causation involves material interactions of energy and does not call upon transcendent entities, operations, or forces that would reinstate a dualistic ontology. It keeps consciousness right where it belongs, in our embodied lived experience. Consciousness is efficacious, not in the sense of causing events, but as contributing to the perception of meanings necessary for intelligent action.

6 Knowing as Transformative Action

In ordinary parlance, to say that you know something is construed either as *knowing that* something is the case—for instance that Djibouti is the capital of Djibouti, that ice melts at around 32 degrees Fahrenheit, and that Immanuel Kant attempted to provide a transcendental argument for the possibility of Newtonian science—or else it is construed as *knowing how* to do something, like riding a bike, eating with chopsticks, installing a door lockset, or titrating a chemical solution. The former type of knowledge is typically called *theoretical*, while the latter is designated as *practical*. When taken in their proper contexts, and relativized to particular goals and values that define different methods of inquiry, these conceptions of knowledge have their place, but only in a *very* limited way.

However, philosophers have far too often slipped into regarding the theory/practice split as rooted in an ultimate metaphysical distinction between a world of fixed realities, which is the object of theoretical knowing, and a world of changing, fleeting temporal events, which is the object of practical knowing. Theoretical knowledge is then taken to be about mentally mirroring the objects and events in a pregiven world, with their distinctive intrinsic properties and relations. Practical knowledge is taken to be about learning how to manipulate material processes in order to realize certain goals or valued ends. This distinction unfortunately tends to be read as two radically different types of knowing, theoretical knowing being a mere contemplation of preexisting realities, while practical knowing is a doing and making in a changing world. Since Dewey regards knowledge as intelligent redirection of experience, he sees *all* knowing as a form of action—as practical. He therefore reinterprets the theory–practice distinction as two different modes of action, one that focuses more on the relatively stable aspects of experience, and the other that focuses on what is more readily changeable through direct action. According to this interpretation, there are not two different kinds of knowing directed toward two different aspects of reality, but a continuum of activities of search, investigation, and reconstruction of nature.

In his 1929 book, *The Quest for Certainty*, which builds on the account of knowing developed earlier in *Experience and Nature*, Dewey analyzed the mistaken motivations and assumptions behind perennial conceptions of absolute knowledge of a supposedly fixed and completed reality. He argued against treating knowing as a metaphorical "seeing" of foundational objects, properties, and relations. Instead, he took knowing to be a form of action in the world—action that changes the world by creating new objects of knowledge. Knowing is an activity by which we transform, through intelligent action, an indeterminate problematic situation into one whose meaning is clarified so that we can move forward in our adaptive behavior.

Dewey naturalizes human intelligence, thereby avoiding any dualisms, such as mind versus body, mental versus physical, thought versus action, knowing versus perceiving, and theory versus practice. He recognizes knowing as a natural process for the transformation of experience. He places the human knower-as-agent *in nature* and actively *transforming nature*—always in the thick of things, trying to grasp the meaning of what is happening, and trying to effect some control over developing circumstances.

The Search for Security through Knowledge

Life is precarious and human well-being is fragile. If we are honest with ourselves, we realize that, despite our best efforts, we often cannot control the vicissitudes of human existence. We go through life in fear and trembling, fearing what may happen, while hoping for the best. Most of us get anxious in the face of an indeterminate or ambiguous situation. We don't handle uncertainty very well (Clark 2016; Hacking 1999; Kahneman 2011; Kahneman et al. 1982; Meyer 1973). We are easily tempted to settle for quick "solutions," in order to dispel our anxiety and doubt, even though these quick fixes may not, in the long run, actually be adequate solutions. It is natural, therefore, and even somewhat necessary, for us to seek stability in a sea of change and indeterminacy. We want a fixed star to guide us on our journey through perilous waters. If only we could have knowledge of what is fixed, unchanging, and ultimately reliable, then, we surmise, *that* would be knowledge most worth having.

There is no such absolute knowledge, no knowledge of eternal forms or ideas capturing the ultimate essences of things in the world (Johnson & Tucker 2021; Putnam 1981; Rorty 1982). There may seem to be a few indubitable truths, such as knowing that "nobody's getting out of here alive," where "here" means this life, but even that chestnut of wisdom is subject to contestation. Even the mathematical and logical propositions that are so often offered up as absolute truths actually depend on extensive sets of assumptions that establish those conceptual systems (and their associated practices), in which such "truths" are taken to be either self-evident or provable (Lakoff & Nunez 2000; Putnam 1981, 2000). There is no systematic body of universal truths upon which to build a superstructure of absolute knowledge. Dewey suggests that only when we give up the quest for certainty, the idea of a fixed and complete universe, and the illusion of unmediated and value-neutral knowledge, are we then able to construct a psychologically realistic account of knowing. That account will see knowing as intelligent transformation of experience—a transformative action, not a beholding of ultimate truths.

The human quest for some measure of stability in life is completely understandable. If all was indeterminate, constantly changing, and chaotic, we could not long survive. We need some predictability, some stability, some reliability of our basic understanding of the meaning of things and events, if we hope to navigate life's troubled waters. However, the conclusion to be drawn from this characterization of the human condition is not that we must therefore seek some allegedly absolute and unchanging bedrock realities, but rather that we should employ methods of inquiry that allow us to adapt to our surroundings, and also to remake those surroundings, in a way that feeds into our survival and flourishing. What we need is a method of intelligent inquiry adequate to our finiteness and fallibility.

There are many good accounts of our desire for certain knowledge that explain why that desire can never be fulfilled, and that provide a realistic alternative naturalistic theory of knowledge (e.g., Churchland 1986; Fesmire 2015; Johnson & Tucker 2021; Kornblith 2014; Putnam 1981; Rorty 1979, 1982; Shook 2000; Weissman 1993). Dewey blames dualistic metaphysical assumptions for our misguided search for certain knowledge. The problem is that we conceive of nature as fixed and completed, with pregiven kinds (concepts or categories) established prior to our investigation into natural events. Then we think of a mind that somehow forms up concepts and propositions that can supposedly mirror mind-independent realities. In this way, what's "inner" (e.g., ideas, thoughts, propositions) supposedly gets connected up with something that's "outer" (e.g., physical objects or states of affairs). A misleading and unnecessary metaphysical dualism creeps in. The *knower* is regarded as ontologically different from what is *known*. The knower is believed to have an individual mind that entertains and manipulates mental objects. We are left with a knowing subject, with its inner thoughts, but without any way to ensure that what that subject has "in mind" actually mirrors external reality. This, in turn, often leads us to seek out unmediated givens (of perception or conception) that might serve as self-validating representations of reality. From those supposedly self-evident building blocks, one then attempts to construct larger knowledge systems that are supposedly immune to doubt.

Dewey calls this the *spectator theory of knowledge*, which posits a knowing self (or ego) that can generate concepts and propositions, and then stand back and assess how well those propositions map onto (or fail to map onto) structures of a mind-independent world. Dewey suggests that many of the problems of knowledge that populate so much of modern philosophy "spring from the assumption that the true and valid object of knowledge is that which has being prior to and independent of the operations of knowing. They spring from the doctrine that knowledge is a grasp or beholding of reality without anything being done to modify its antecedent state—the doctrine which is the source of the separation of knowledge from practical activity" (Dewey 1929/1984, 157). Theory, from the Greek *theoria*, is looking at, viewing, or beholding something, as when an audience beholds a dramatic performance. Contemplation of this sort involves a spectator who sees and appreciates what is going on in the play, but does not typically participate in the play itself. So, the spectator-knower remains separate from what it knows, or beholds, and it plays no role in constituting what is known. It merely appropriates and processes what is given, either to sense perception or to thought.

Knowing as Intelligent Remaking of Experience

Dewey's alternative, nondualistic, naturalistic account treats knowing as a form of activity in the natural world: "If we see that knowing is not the act of an outside spectator but of a participator inside the natural and social scene, then the true object of knowledge resides in the consequences of directed action. . . . For on this basis there will be as many kinds of known objects as there are kinds of effectively conducted operations of inquiry which result in the consequences intended" (1929/1984, 157).

Dewey appropriated the term "intelligence" for his distinctive conception of the process of knowing, according to which "known objects exist as the consequences of directed operations, not because of conformity of thought or observation with something antecedent. We may . . . give the name intelligence to these directed operations. Using this term, we may say that the worth of any object that lays claim to being an object of knowledge is dependent upon the intelligence employed in reaching it. . . . Intelligence means operations actually performed in the modification of conditions, including all the guidance that is given by means of ideas, both direct and symbolic" (1929/1984, 160). In short, *knowing is the operation of intelligence in remaking experience in light of our interests. values, and environments, so as to maintain and enhance our viability*. What we count as knowledge will depend on the methods we employ to reconstruct experience. Based on past experience, we project expectancies about what we will encounter in experience under certain conditions, and then we test out whether, and to what extent, those expectancies are confirmed in the resulting experience. The crucial difference between knowing as contemplative seeing and knowing as experience-transforming inquiry "is that between a mind which beholds or grasps objects from outside the world of things, physical and social, and one which is a participant, interacting with other things and knowing them provided the interaction is regulated in a definable way" (Dewey 1929/1984, 160).

Dewey published *The Quest for Certainty* only a couple of years after Werner Heisenberg's revolutionary work (1925, 1958) on quantum uncertainty burst on the scene in the late 1920s. Dewey saw in Heisenberg's research a validation of Dewey's idea that the mode of inquiry selected shapes both what is knowable and how it is known. The act of inquiry determines to a certain extent the object known. Heisenberg challenged the reigning Newtonian causal determinism, which held that any object or particle could be definitively identified by the combination of its location and velocity (momentum) vector in an infinite expanse of space (and time). In a Newtonian universe, if we could specify the precise location and velocity of all particles of matter in a present moment, then, in principle, we should be able to predict with certainty future states of affairs. Heisenberg's experiments with light particles showed, to the contrary, that if we were to fix one variable (say, location), then the other variable (momentum) can only be determined with a degree of probability (but not with certainty). There are thus fundamental indeterminacies *in nature*! This indeterminacy is an objective metaphysical fact about the nature of matter, and not merely a limitation of human cognition. There is a limit to the precision with which we can determine pairs of physical properties of any particle.

One significant corollary of this hypothesis is that scientific laws (i.e., causal laws of nature) cannot be universal statements of rigid causal uniformities of natural events. They do not define a causally deterministic system governing all natural goings-on. Instead, causal laws are "*formulae for the prediction of the probability of an observable occurrence*. They are designations of relations sufficiently stable to allow of the occurrence of forecasts of individualized situations" (Dewey 1929/1984, 165; italics in original). Laws are thus tools for helping us predict the relative probability of certain natural occurrences (Cartwright 1983; Peirce 1878, 1898/1992). They help us understand the meaning of things and events as they issue in experiences, but they make these predictions concerning anticipated things and events only with a certain degree of probability, not with absolute certainty. A scientific law of nature is thus only of use just so long as it actually does give us insight into the meaning of events and allows us to make somewhat accurate predictions and to manage our present situation.

Another important aspect of Heisenberg's views is that the observer can actually affect the character of what is observed. The act of inquiry influences the object known: "The particle observed does not *have* fixed position or velocity, for it is changing all the time because of interaction: specifically, in this case, interaction with the act of observing, or more strictly, with the conditions under which an observation is possible" (Dewey 1929/1984, 162). It may seem strange to say that an act of observation actually changes the nature of what is observed, so that the activity of inquiry changes the nature of what is known, but this becomes less awkward when it is seen that what we take as knowledge will depend on the values, purposes, and goals shaping our present inquiry, and thereby determining what should count as a satisfactory explanation (Hanson 1958). Knowing is a form of doing that changes the object known by giving it new meaning.

The spectator theory of knowledge must be abandoned. We have no value-neutral theoretical standpoint on a mind-independent world. We are not separated from what we are attempting to know, but rather are in and of natural processes that include our activity of knowing—what is known is seen to be a product in which the act of observation plays a necessary role: "Knowing is seen to be a participant in what is finally known. Moreover, the metaphysics of existence as something fixed and therefore capable of literally exact mathematical description and prediction is undermined. Knowing is, . . . a case of specially directed activity instead of something isolated from practice. The quest for certainty by means of exact possession in mind of immutable reality is exchanged for search for security by means of active control of the changing course of events" (Dewey 1929/1984, 163). Acts of knowing are validated according to how well they anticipate forthcoming events, based on understanding the meaning of present happenings. The process of projecting expectancies and testing them against emerging experience is an ongoing process. It consists of repeated projection, projection, and more projection, interspersed with continual testing, testing, and more testing, with adjustments being made in response to failed expectations and changing conditions.

We are *in* and *of* what we know, and in the most visceral, engaged way. Knowing is a way of intelligently exploring the meaning of our situation, and then reworking it in ways that permit us to achieve greater viability and a measure of well-being. Knowing *is not*, and cannot be, the achievement of a God's-Eye point of view on a fixed world (Putnam 1981). We cannot escape our perspectival and active relation to what is investigated and known. All knowing is motivated and rooted in exploratory activity as we probe our world, guided by our most basic homeostatic and allostatic needs.

Knowing is thus a term of achievement. We can be said to know something only when, having encountered a problematic situation, we have initiated a value-directed and motivated search that results in a relative satisfaction of the needs and interests driving our inquiry. We are not disengaged spectators of a mind-independent world, but rather makers and doers, at work reconstructing experience, which is to say, remaking our world. According to Dewey, then, "knowing is one kind of interaction which goes on with the world. Knowing marks the conversion of undirected changes into changes directed toward an intended conclusion" (1929/1984, 163). Knowers are agents, actors, and doers engaged in testing out projected expectancies about what will be encountered in experience. The only relative confirmation and validation available to us comes from testing our projected expectancies against what we encounter at the interface with our physical and social surroundings.

Nature Is Not Inherently Rational

Dewey's account of knowing as reconstructive action helps us evaluate the perennial idea that we live in a rational universe, in which our rational mind is able to discern the ultimate rational structure of the world. On this view, we can know the world because it exhibits the very rationality (as structure) that is present in human minds. This version of rationalism is false. We are not possessed of absolute Reason or Understanding (*nous*) that is identical with the rational structure of the world (*logos* as a manifestation of *nous*). The cosmos can be said to be rational only because our intelligence is part of what makes it what it is and shapes its development. The import of this can be illustrated with a couple of analogies. As you drive, on narrow two-lane blacktop roads, through the extensive lava fields of the high desert of central Oregon, you might jokingly marvel at the fact that, in some places, the lava flow has seen fit to stop *right on the edge of the road*! How considerate of it to do so. It may therefore seem as if there is some intrinsic relation between lava flows and roads in that part of Oregon. Well, there *is* a relation, but it stems from the fact that the roads were built, by humans with certain interests, to get us around the lava flows, so it is no miracle that they circumvent the flows so closely. Dewey makes the same point with his example of how cities often seem to be near rivers. But there is no intrinsic relation of cities and rivers. The reason for this is just that humans built those cities by rivers precisely to satisfy their need for farm irrigation, drinking water, transportation of goods, and recreational activities.

Dewey employs the city-near-river analogy to explain why, just as there is no intrinsic relation between cities and rivers, so also our natural world is not intrinsically or essentially rational. Nature exhibits rationality just where humans have labored to organize natural processes to help them satisfy their basic needs and valued interests. Nature is not rational *in itself*; instead, it is capable of being intelligently engaged just insofar as we have developed in and through ongoing interaction with our natural and social surroundings. In other words, nature *becomes* somewhat rational as a result of our adaptive engagement with it. Dewey concludes, "Abandon completely the notion that nature *ought* to conform to a certain definition, and nature intrinsically is neither rational nor irrational. Apart from the use made of it in knowing, it exists in a dimension irrelevant to either attribution. . . . Nature is intellig*ible* and understand*able*" (1929/1984, 168). Nature is intelligible and understandable just to the extent that we can intelligently interact with it to direct the course of events to some degree. However, achieving such understanding is not a matter of discerning the ultimate rationality of nature. Such rationalism is a remainder left over from the theological notion of a world created by a rational God, thereby manifesting an ultimate knowability based on Divine creation and purposiveness.

Abandoning the idea of a fixed and completed block universe requires jettisoning the correlative idea of fixed and completed objects of knowledge that are supposedly beheld by a knowing ego. However, the alternative to this view is not everyone's cup of tea. That alternative is the view that objects known are constructed, instead of given. They are the result of action in the world, rather than a precondition of it. The attempt to know something requires an inquiry into its meaning, which to say, its generating conditions and its implications for experience. The inquiry involved is motivated by our visceral needs, guided by our emotional responses, and engaged in the reworking of experience toward some end-in-view. The meaning of some experienced object or event is the result of our mode of inquiry and not a precondition for that inquiry. Dewey summarizes: "The organs, instrumentalities and operations of knowing are inside nature not outside; Hence they are changes of what previously existed: the object of knowledge is a constructed, existentially produced, object" (1929/1984, 168). Objects are *events with meaning* (Dewey 1925/1984). Creating the object known is based on recognition and enactment of new meanings for a thing or event. The known object is thus an achievement of a process of inquiry. This is the sense in which an object known is produced by acts of inquiry: the object changes as its meaning (its implications for experience) changes. To know that object is to know how you can interact with it in various physical, social, and cultural activities. It is to understand how that object stands in relation to the system of shared meanings that is mind.

Knowing as Reconstructive Activity

What is revolutionary about this view of knowing is that it overthrows the spectator theory of knowing and replaces it with an account of the knowing process as embodied, action-oriented, and transformative of experience. This view "installs man, thinking man, within nature" (Dewey 1929/1984, 168). We cease to see ourselves as outside nature, looking in on it in search of indubitable perceptions, concepts, or propositions concerning how things are in nature. Instead, the knowing agent is always embedded in developing natural events, in which that natural agent shapes the development of experience through its bodily immersion in natural processes. There is no radical separation of mind from body, or self from world. There is only the ongoing flow of natural processes, some of which involve agents reworking their understanding of nature by continually testing out the meaning of things and events.

As inextricably embedded in natural events, the test of intelligent inquiry is how well it responds to imminent needs, solves problems by reducing indeterminacy, and shows us paths for enriched and liberated action. Intelligence is Dewey's preferred word for the experience-transforming activities of motivated inquiry into nature. "A man is intelligent not in virtue of having reason which grasps first and indemonstrable truths about fixed principles, in order to reason deductively from them to the particulars which they govern, but in virtue of his capacity to estimate the possibilities of a situation and to act in accordance with his estimate. . . . Wherever intelligence operates, things are judged in their capacity of signs of other things. If scientific knowledge enables us to estimate more accurately the worth of things as signs, we can afford to exchange a loss of theoretical certitude for a gain in practical judgment" (1929/1984, 170).

In giving up the quest for certainty, we lose nothing, because we never did have access to certain, indubitable knowledge. What we gain is psychologically realistic practical wisdom, the intelligence for interrogating nature and finding our appropriate place within it, thereby enhancing our viability and well-being. Practical wisdom of this sort is not a second-best concession to inferior understanding, but rather our best opportunity to explore the meaning of our world through directed activities that reconstruct experience relative to our needs and interests.

A naturalistic theory of knowing has to see the knower *in* nature, not above it. A naturalistic theory has to see knowing as a remaking of experience, not a beholding of a pregiven state of affairs. It situates the knower in the middle of natural happenings, searching its way through those events and shaping the course of development of those events. "When an interaction intervenes which directs the course of change, the scene of natural interaction has a new quality and dimension. This added type of interaction is intelligence. The intelligent activity of man is not something brought to bear upon nature from without; it is nature realizing its own potentialities in behalf of a fuller and richer issue of events" (Dewey 1929/1984, 171). What we now need to examine, therefore, is what this transformative process of knowing consists of, as a method of inquiry.

Knowing as Need–Search–Satisfaction

Dewey was keen on naturalizing epistemology, something that would later become a cottage industry (Kornblith 1980, 2014; Quine 1969). In what he called in *Experience and Nature* "the method of empirical naturalism," Dewey emphasized the actual knowing process, as bootstrapping, correcting, trying to make sense of one's situation, always tied to some form of human action, real or imagined.

The emphasis is on human experience, the gateway of human engagement and understanding. The empirical method is broad in scope and deep in meaning; not a barren landscape of epistemic abstraction, codification, and formalism. Our knowing is not a matter of bringing an external perspective or conceptual framework to bear on a pregiven reality. Instead, knowing is embedded in a developing transformation of experience in the most visceral and intimate fashion. Knowledge requires a practical engagement with processes of experience, rooted in habits of continuous adaptive functioning.

Acts of knowing appropriate the need–search–satisfaction process that drives our biological functioning and motivation (Johnson & Schulkin 2020; Johnson & Tucker 2021). The biological process of need–search–satisfaction is the basic structure of embodied allostatic development at the level of basic life maintenance, and it is also a precursor of, and condition for, our higher acts of knowing. In this way, intelligence is naturalized through and through, always operating within experience, not outside it.

Knowing is the endless clearing away of obstacles that impede our interests, projects, and epistemic urges. Within an indeterminate, problematic situation, a need gives rise to a search for a way to resolve the experienced ambiguity or indeterminacy, in order to re-establish a satisfactory adaptive behavior. This ever-recurring need to remove the irritation of doubt that frustrates fluid action is a core theme for pragmatists like Dewey and

Peirce (1868, 1878). The obstacles generate the motivation and energy for solutions, solutions to quell the unease and discomfort by reorganizing experience.

In the "The Supremacy of Method" chapter of *The Quest for Certainty*, and also in *Experience and Nature*, Dewey provides a thorough naturalization of basic bio-behavioral organizing principles that underlie our higher knowing activities. At the level of biological functioning, *need* is realized as a disequilibrium in the homeostatic life processes of an animal. This initiates a *search* process, which attempts to recover the lost equilibrium (homeostasis) or to establish a novel dynamic equilibrium (allostasis). This is achieved through activities of exploration and transformation of the environment, in response to arising needs. When this search phase is successful, the result is temporary satisfaction. This is a recurring process of animal life, an ongoing series of cycles of falling out of harmony with one's surroundings and acting to re-establish some measure of reintegration and equilibrium. A simple example in the animal kingdom is the case of a biological need for nutrition, as necessary for maintaining or regaining the body's life-sustaining functions. In animals of sufficient complexity, their brain sends chemical and electrical signals that, along with the secretion of hormones throughout the brain and body, affect motivational systems that activate various sensory and motor functions, leading the animal to forage for food.

In most nonhuman animals, this is pretty much an automated, habitual, nonconscious process of changes in the animal's body-state that generate search behaviors. For these animals, "to live signifies that a connected continuity of acts is effected in which preceding ones prepare the conditions under which later ones occur" (Dewey 1929/1984, 179). In humans, however, there is something *more*, and that more is mind (in Dewey's broad sense), which makes a big difference, in two ways. First, as the complexity of animals increases through evolution, their environments become more complex and afford more possibilities for interaction and appropriation. The downside of this is that our more complex environments become more indeterminate and potentially problematic: "Conditions of the environment become more ambivalent; it is more uncertain what sort of action they call for in the interests of living" (Dewey 1929/1984, 179). However, there is a big upside of this increasing complexity and indeterminacy of the animal–environment interaction; namely, there are more opportunities for possible action within a particular environment in fulfillment of needs.

For humans, with the emergence of mind, problematic situations acquire a mental quality of doubtfulness, which spurs us to actions geared toward the restoration of allostasis, thus reducing the indeterminacy of the situation. In this way, the need–search–satisfaction process becomes intellectual and cognitive, as we engage in thought as a way "to locate the nature of the trouble and form an idea of how it may be dealt with—so that operations may be directed in view of an intended solution" (Dewey 1929/1984, 181).

The emergence of mind, as a cultural system of developing meanings, is a monumental game changer. In humans, and perhaps to a certain degree in a few other animal species, conditions arise that make possible intelligent, reflective redirection of experience in service of our needs and interests. What was before only a biological process of need–search–satisfaction now becomes an intellectual process, one that recruits the same need–search–satisfaction structure to aid the intelligent reconstruction of experience in the service of desired ends. For humans, a problematic situation has an immediate *emotional* quality, that of anxiety and doubt spawned by indeterminacy. It also has a *volitional* quality, realized as "the tendency of the reaction to modify indeterminate, ambiguous conditions in the direction of a preferred and favored outcome, to actualize one of its possibilities rather than another" (Dewey 1929/1984, 180). This process becomes *intellectual* in character, as we reflectively examine the nature (i.e., the meaning) of the problem and project possible solutions. *There is thus a fundamental continuity between the need–search–satisfaction process as it operates at the biological level, geared toward animal viability, and its operation at the level of reflective search, deliberation, and choice.* In other words, the need–search–satisfaction pattern of organic functioning is appropriated for thoughtful problem solving. Via this basic continuity, all knowing remains bodily, emotional, value-based, and action oriented, from primitive animal search operations all the way up to our highest achievements of scientific and artistic reasoning.

The sense of irritation and unease that characterizes a doubtful, indeterminate situation is rooted in the biology of the organism. It is all about the recurring phases of appetitive search and consummatory satisfaction, as a means to survival and well-being. These phases are shaped by basic bio-behavioral systems, temperament, and capabilities (Craig 1918; Tinbergen 1951, 1963). This includes especially the search for food, water, salt, shelter, safety, and sexual access. It also includes the demands of reproduction, managing of offspring, attachment to others, and social capabilities necessary for cooperative activities. Diverse chemical modulators, such as peptides, neuropeptides, steroids, and neurotransmitters across both the brain and the major body and organ systems are involved, for both vertebrates and invertebrates alike, with their diverse kinds of brains and bodily capabilities (Fitzsimmons 1998; Herbert 1993; Herbert & Schulkin 2002; Schulkin 1999; W. J. Smith 1997).

As we saw earlier, all these activities are carried out in the context of suitable environments where procurements are possible and central states of the brain are linked to environmental niches. They are organized temporally via endogenous generators tied to clock-locked activities (e.g., twenty-four-hour, monthly, and seasonal cycles) (Moore-Ede 1986; Moore-Ede et al. 1992; Mrosovsky 1990; Nelson et al. 2002; Shettleworth 1998). They are tied into environments, both internal and external, that supply resources relative to the appetitive and consummatory events (i.e., search leading to satisfaction).

Foraging for coherence is the dominant theme. Various appetitive and consummatory events involve diverse learning capabilities geared toward enhanced viability. For humans and other animals this involves others, so social coordination and adjustments are omnipresent. We learn how to respond to salient events and to the incentives that are alluring. We learn cooperative interpersonal behaviors for coordinating actions in pursuit of shared interests and goals. We are continually exploring, sampling, learning, and assessing the value of various resources, relative to our well-being and that of others. Importantly, diverse learning essential for continued functioning is not dependent upon a current need state, insofar we learn and remember where resources are to be found and how to procure them, whether we are in the appetitive search phase or not (e.g., latent learning, sensory preconditioning).

Information is thus not a set of abstract propositions, but rather vectors of interaction; we search for information, mostly without conscious effort, but sometimes we probe and strengthen our epistemic urges to enhance our search activities. Informational deprivation, imagined or not, is a key component in what moves us to act. But the epistemic drive to fill in information, to quell visceral reminders of this need, is operating from the onset and continues throughout the life cycle. Inference and expectations are running themes in the organization of action (Loewenstein 1996). The epistemic orientation, whether by necessity—because survival depends on it—or simply musing about objects of interest in the sense that Peirce (1898/1922) adumbrated, is a key dimension of the naturalization of the knowing process and the continual growth of our flexible, expanding choice and freedom earned in action (see also Dennett 1984; Neville 1974; Schulkin 2007).

As animals in our natural and social worlds, we are guided by heuristics we come with in our innate orientations and also acquire via continual learning. The root metaphor for these epistemic orientations is not that of a blank slate or a blooming buzzing confusion, but rather a self- and world-transforming animal—an artist of experience. Alleged divisions between innate predilections and diverse forms of learning are diluted, insofar as our epistemic inquiries are wrapped up in our search for both animal viability and human meaning and understanding. Dewey just makes the whole thing empirical, or at least part of the empirical process of self-correction and test, issuing in attempted control where possible within "a constant mixture of the precarious and the stable" (Dewey 1925/1981, 243).

The environment is not other to us; we are in it, part of it, transforming it by our actions. The organism is part of the natural world. There is no mind/body split, and there is no isolation of the knower from the known. There is just the difficulty of temporarily resolving problematic situations and attaining satisfaction. Objects are not reified into Platonic forms, products of Kantian pure organizing structures of a transcendent ego, or results of passive association, as envisioned by Hume. Objects are events with meaning, where meaning arises in and through our bodily, motivated, and value-based engagement with our surroundings.

Two important implications of this perspective are (1) that all knowing is embodied, motivated, and based on biological and social values, and (2) that there is a continuum between the need–search–satisfaction process in biological functioning and that same structure of phases in higher-level conceptual and propositional modes of knowing. It is this continuity from biological to higher epistemic practices that underlies naturalism and allows us to avoid metaphysical and epistemic dualisms. Mind and thought are entirely natural processes, and any separation of mind from what is thought or known is relative and permeable. Dewey concludes: "Thinking is the actual transition from the problematic to the secure, as far as that is intentionally guided. There is no separate 'mind' gifted in and of itself with a faculty of thought; such a conception of thought ends in postulating the mystery of a power outside of nature and yet able to intervene within it. Thinking is objectively discoverable as that mode of serial responsive behavior to a problematic situation in which transition to the relatively settled and clear is effected" (1929/1984, 181).

What Is Knowledge?

So, what is knowledge? It is a term of achievement applied when an operation of inquiry, both as biological foraging and intelligent reflective thought, leads from an unsettled situation to a reconstructed and more clarified and meaningful experience, which allows us to function, not perfectly, but *well enough*: "knowledge or a known object, marks a question answered, a difficulty disposed of, a confusion cleared up, an inconsistency reduced to coherence, a perplexity mastered. . . . Knowledge is the completed resolution of the inherently indeterminate or doubtful" (Dewey 1929/1984, 181).

Dewey is sometimes mistakenly attributed with the view that knowledge occurs whenever an individual subject's doubts are resolved and that person regains a sense of certitude. This would amount to saying that the criterion of knowledge is the calmness of an individual mind. Nothing could be farther from the truth. If the criterion of knowledge was absence in an individual person of any anxiety or doubt, then the most prejudiced, ill-informed, unreflective lout would be the exemplar of all knowing. Instead, Dewey's view is that the indeterminacy that constitutes a problem resides *in the situation*, and not merely in the mind of an individual who is confronting that situation. Therefore, resolution of indeterminacy is an objective process, issuing in a transformed situation, rather than a mere psychological readjustment: "The commonest fallacy is to suppose that since the state of doubt is accompanied by a feeling of uncertainty, knowledge arises when this feeling gives way to some assurance. Thinking then ceases to be an effort to effect change in the objective situation and is replaced by various devices which generate a change in feeling or 'consciousness'" (Dewey 1929/1984, 181).

This subjectivizing of a knowing state explains the human tendency to replace feelings of doubt and anxiety with anything that will quell them, and as quickly as possible. We do not well abide doubt and the anxiety generated by indeterminacy. The unease spawned by indeterminacy is uncomfortable. Too often, we take the easy way out. We try to make those feelings of anxiety go away by jumping to a conclusion without sufficient inquiry, by oversimplifying, or by ignoring facts. These ploys may provide a temporary respite from doubt, but it can only be temporary, because, while our psychological dissonance may be reduced, the indeterminacy of the *situation* has not been resolved. This tends to lead to repression, sublimation, and other compensatory psychological strategies. Dewey concludes: "If doubt and indeterminateness were wholly within the mind—whatever that may signify—purely mental processes ought to get rid of them. But experimental procedure signifies that actual alteration of an external situation is necessary to effect the conversion. A *situation* undergoes, through operations directed by thought, transition from problematic to settled, from internal discontinuity to coherency and organization" (1929/1984, 185).

The remedy to a problematic situation is inquiry—sustained, intelligent, reflective, critical inquiry. As emphasized earlier, the need is for more projection of expectancies, more foraging for possibilities, more controlled testing of hypotheses, and more self-critical reflection. In their consummatory form, these characteristics define scientific method at its best. So-called scientific method, which is really multiple methods, is just an exemplary instance of the empirical denotative method described in chapter 2, which is the basic method of empirical inquiry (Johnson & Tucker 2021).

Dewey understood empirical method as quite broad in scope. What he called "the supremacy of method" (1929/1984) is the grounding of human inquiry in the everyday fabric of human meaning and understanding, and appropriating that process for forms of higher-level inquiry (Heelan & Schulkin 1998). Science, like art, roots in our natural dispositions for exploration, discovery, understanding, and resolution of concerns.

Within this orientation, there is no one method of inquiry. There are many, depending upon the context, the tools, and the issues of focal concern (Kuhn 1962). What matters is the evaluation of evidence for a given assertion, based on testing, attempted replication, and expansion of meaning. Nothing is abstract about this. It is embodied, engaged knowing that requires context, cultivation of capabilities, and sufficient resources.

The history of experimental method takes us from the accepted value of the honesty and authority of the investigator, whose assertion is supported by what the investigator claims to have evidence for, to an emphasis on what others can confirm with publicly recognized methods for discovery of evidence. The difference between Thomas Hobbes and Robert Boyle on this issue was the development of the experimental methods that Boyle introduced in his study of the pumps in the heart (Shapin 1995; Shapin & Schaffer 1985). Both Hobbes and Boyle were empirically oriented, but Boyle emphasized the development of methods and routes of demonstration, followed by replication of relevant experiments by others.

It was this empirical, experimental sensibility that Dewey raised to a crescendo of meaning-making and inquiry present in all aspects of life, from basic allostatic regulation all the way up to scientific experimentation and discovery. Evidence, for Dewey, was never grounded in what the logical empiricist philosophers (e.g., Mach, Carnap, Moore, Ayer) called "sense data," which were taken to be basic uninterpreted sensory givens that were supposed to provide a bedrock for knowledge. Pragmatism replaced claims about atomistic sensory givens with the focus on shared methods of confirmation and disconfirmation. Because absolute confirmation was recognized to be impossible (Hempel 1965), and even resistance to falsification was not a guarantee of truth (Kuhn 1962), pragmatists emphasized the organization of research programs and the development of fruitful forms of inquiry and discovery that could be employed by a community of inquirers. What is emphasized are fruitful ways of gathering evidence, making sense of things, and testing hypotheses. The hands of inquiry are dirty hands, hands actively probing and working the soil of experience. There can be just as much embodied inquiry and problem solving in nuclear physics as in work in a garden or building of a house. Knowing is engaged problem solving, a completely natural process, whether in acts of physical exploration and building, or in acts of intellectual hypothesis formation and testing. The evolutionary emergence of mind allowed us to extend, by leaps and bounds, the cultivation of the diverse forms of self-corrective reflective activity.

The basic challenge for the sciences is to provide methodological tools to resolve our existential unease by enriching our understanding of our world. Science is thus a natural outgrowth of human learning, building, theorizing, and testing. It replaces unreflective, habitual reaction to events with adaptation to the environment, change of the environment, knowing the terrain, and enacting the better, rather than the worse, within a given situation.

However, to many folks, advanced scientific research can often seem about as counterintuitive as anything could be. For example, the notion of action-at-a-distance is anything but intuitive. The idea of one thing causing or affecting another thing that it doesn't even make direct contact with seems impossible. Moreover, vast stretches of abstract mathematical physics traffic in notions that make little or no sense to nonscientists. What could be more counterintuitive than quantum indeterminacy, where, for instance, light is taken either to be a wave phenomenon or a series of particles, depending on which experimental designs are used?

Robert McCauley (2011) has explored the ways in which advanced scientific research often goes far beyond our everyday, commonplace modes of understanding and inquiry, and how it can sometimes traffic in highly counterintuitive phenomena and concepts. He calls this the "unnaturalness of science," which he contrasts with the naturalness of religious tendencies to see agency everywhere and to explain events as the result of activities of special kinds of agents. The character of various scientific modes of inquiry and explanation makes it hard for some people to embrace science over religious explanations built on extensive postulation of spiritual agencies. McCauley suggests that the relative "unnaturalness" of science should urge us to carefully cultivate and nurture methods of empirical inquiry and explanation, whereas religious accounts will tend to take care of themselves. McCauley also observes that advanced scientific method is a refinement and extension of empirical strategies of inquiry that operate both at the biological and higher cognitive levels of inquiry: "Science calls on some cognitive processes that come to humans relatively easily. Humans are naturals at thinking up theories, and they are sensitive to some kinds of evidence that bear on those theories' truth. Gopnik, Meltzoff, and Kuhl review evidence that indicates that these proclivities are already in place and developing while babies are still in the crib" (McCauley 2011, 105).

Dewey's claim about the continuity of ordinary, day-to-day modes of empirical inquiry with those of advanced scientific research is that both employ the need–search–satisfaction phases of inquiry that first arise in our biological functioning, and both make extensive use of body-based meaning and thought. Dewey's notion of method is experimental, where "experiments" can be as simple as testing to see whether you are allergic to poison oak or whether your laundry detergent "gets the stains out" (as the TV add promises), or it can be as complex as the use of massive technologies to test whether light is curved as it passes a large gravitational body, or to assess the behavior of subatomic particles using the Hadron Collider. The sciences are thus experimentalism on steroids. The point is that the method employed is essentially the same in structure for these more mundane inquiries, as well as for our most spectacular scientific research. McCauley observes the key role of ongoing criticism, hypothesizing experiential predictions, testing, retesting, and theory revision, which are the hallmarks of good science, but which are often not well represented in ordinary day-to-day inquiry (2011, 118–137).

Science is a precious tool for the anticipation of events and for understanding their consequences. Peirce was the only experimentalist of the early pragmatists (Peirce & Jastrow 1884), but Dewey was a graduate student at Hopkins when Peirce was on the faculty. Peirce set up the first experimental laboratory in psychophysics in the United States (Cadwallader 1975) and went on to outline the nature of experimental inquiry in a detailed manner in journals like *Popular Mechanics*. Dewey took note and realized the importance of empirical inquiry in all dimensions of human experience.

Dewey made science—empirical method—plain for ordinary folks like us, not the high-level specialists that Peirce so well understood and regarded as his audience. Dewey emphasized pedagogy and the need to expand the self-corrective orientation to problem solving generally, in all corners of our daily lives. Science broadly understood is the extension of our commonplace empirical orientation to problem solving. Indeed, we come into the world, early on in ontogeny, prepared to become inquirers (Gopnik & Meltzoff 1997). As we develop, we learn to project expectancies about what we will experience and then to explore whether those expectations are confirmed at the sensory–motor interface with our surroundings. We engage in a form of Bayesian statistical inference, in which, based on past experience, we project expectancies for future events and then explore whether they are supported by evidence that can be replicated by others (Baron 1988; Clark 2013; Gigerenzer 2000; Hohwy 2013). In this way, through empirical inquiry, meaning grows.

Statistical inference is as commonplace to our minds as is liking sweet foods. It comes with the territory, the territory of human inquiry and knowing. We project expectations, calculate, probe our surroundings, recalibrate expectations, and act to transform experience and to generate new meanings. Science is just the natural extension of this process in communities of like-minded inquirers, but codified and more highly developed via tools, technologies, and impressive abstractions and generalizations. In this way, the evidence for claims to knowledge is put in a larger ecological/social context with more elaborate and refined procedures for analysis and evaluation of hypotheses.

Dewey concludes that "the fundamental advantage of framing our account of the organs and processes of knowing on the pattern of what occurs in experimental inquiry is that nothing is introduced save what is objective and is accessible to examination and report" (1929/1984, 183). Science trades away a (nonexistent) infallibility for fallible, but correctable, projection and continual testing of hypotheses, often within a community of inquirers. The illusory certainty that is lost is replaced by reliable empirical methods whose results are subject to critical assessment by others, under the guidance of shared standards of testing. The occasional need to correct prior hypotheses is not a failure of science, but a vindication of critical empirical methodology.

Knowing, in this embodied sense, is an ongoing existential process of life management. Pedagogy is lifelong, as the bodily responses and bio-behavioral orientations are integrated into doing stuff, building stuff, inventing stuff. Dewey sees empirical inquiry as attentive to and shaped by events that matter—that pose threats, offer rewards, or pique our curiosity. This curiosity gets turned toward eliminating problems and opening up new ventures of knowing, but now with some measure of reflective control over what happens. Our natural predilection is not to a blooming confusion, but to orientation to resolve dilemmas and to seek resources required to remain viable. This need to know is in our deepest interest, at the very heart of our evolution. We became the pedagogic species that we are by continuing learning over the life cycle, and not merely in early development. It is this pedagogical process that Dewey is tracking, cultivating, and making the cultural fabric of inquiry that is continuous with some of our deepest natural predilections. From disruption of reaching a solution to a problem to wanting to know about our world, we are embedded in and expanding our horizons. This, in turn, gives us pleasure, satisfaction, and growth of capabilities for knowing. Epistemic urges are up close and profoundly personal. Our bodily growth, emboldened by continuous cultivation of neural capability and bio-behavioral solutions brings satisfaction and enjoyment in the knowing process.

The naturalistic account of mind and knowing that we are developing here is rooted in the continuity of biological processes and forms of thought and inquiry. There is no dualism whatsoever. Instead, structures and processes of bodily search and satisfaction are appropriated for more complex cognition and problem solving. All of this knowing activity operates always within the natural processes of experience: "The organism is a part of the natural world, its interactions with it are genuine additive phenomena. When, with the development of symbols, also a natural occurrence, these interactions are directed towards anticipated consequences, they gain the quality of intelligence, and knowledge accrues. Problematic situations when they are resolved then gain the meaning of all the relations which the operations of thought have defined" (Dewey 1929/1984, 187). We remain always within the thick of things, and our search for viability, sometimes issuing in knowledge, is a remaking of things and events within experience, using the resources of our natural embeddedness in the world.

7 The Making of a Self

Your *self* is *you*! Whenever you ask who somebody is, attribute moral responsibility to them, or wonder whether they've changed over time, you are concerned with their selfhood, their identity, what makes them who they are. Selfhood is therefore a very significant topic for every human being, or should be. It's a matter of who you are, and who you might aspire to be. So, who *are* you? Are you an eternal soul, who happens to hang out temporarily in a particular body during the course of your earthly life, only to pass over into some eternal realm as you slough off your mortal coil? Or do you see yourself as just a somewhat sophisticated animal—the current product of the evolutionary heritage of your species and your individual development over your life span—whose self as presently formed will cease to exist upon your biological death? Or, maybe you think the very notion of a unified substantial physical or spiritual self with a core identity is a profound illusion, and that you are only a series of brief moments of what Buddhists call dependent arising in the ongoing processes of nature, but without any fixed identity.

We want to suggest that self-identity is an important phenomenon that is meaningful to us and that does genuine work, even though the self is not a "thing." You might wonder why we even need a chapter on the self, since we already have chapters on mind and consciousness. The short answer is that the self is neither just mind, nor just consciousness, but rather both, and more. First, there is more to the self than mind. Mind is a vast cultural system of shared meanings, practices, and values that individual selves can appropriate and participate in, whereas the self is also rooted in, though not reducible to, the body. There is more to your self-identity than the cultural meanings you draw upon for your self-understanding. For one, there is your lived body.

Second, the self is not reducible to consciousness, or to a conscious ego. It cannot be the sum total of your conscious states, for the obvious reason that the vast majority of our so-called cognition goes on beneath our conscious awareness (Damasio 1999, 2010; Helmholtz 1873; James 1890; Rozin 1976). Your selfhood is enmeshed in all of the nonconscious processes of searching, approaching, avoiding, moving, manipulating, and valuing tied to your bodily perceptual and motor capacities, your emotions, and your motivational systems. Body-based probing and transformation of the world is a recurring theme throughout the life cycle, from the most mundane to the most elaborate operations of self-formation. Consequently, we are not even aware of most of what makes us who we are, at a given moment in time.

If the self, then, is not reducible only to mind, to consciousness, to your flesh and blood body, or to some part of the environment, then what, and where, is it? The answer is that selfhood is based in all four of these dimensions of experience, and the key issue thus becomes how these four aspects interweave to constitute your selfhood. From a naturalistic perspective, we must start with the recognition that there is no self without a body, even though you are more than your body *simpliciter*. Following Damasio, we will begin the discussion with the self as a combination and intertwining of what he calls the protoself, the core self, and the autobiographical or extended self. Following Dewey, we will see the self as a dynamically developing interpenetration of habits of perception, thought, valuing, and action. Because habits incorporate aspects of the environment, the self cannot be locked up within some alleged inner mental realm but must be realized through our ongoing organismic interactions with our surroundings. Neither can the self be locked up entirely in the external world or in cultural values and practices because the habits that form our identity require a functioning body and brain for their realization.

At any particular moment, the self as presently formed results from the interrelations of prior habits, but the self is also always in the making, due to responsiveness to newly emerging conditions that call forth newly formed habits (Mead 1928/1972, 1934). In this sense, the self is always a self-in-process, an intertwining of patterns of organism–environment interactions that are tentative and transitory. The origin of all of this self-making turns out to be our bodies—not simply the body as a bag of flesh, but rather the body-in-the-world, the lived body, the body that is the organic medium of our perceiving, feeling, thinking, valuing, and doing (Merleau-Ponty 1962). Then, through engagement with others within one's particular cultural system, the self is further built up as it acquires meanings and cultural narratives that give definition to the self. The issue, fundamentally, is what kind of self is in the making, and who we aspire to be, such as our better angels, faint and frail.

Damasio's Blending of Self Processes

In chapter 5 we discussed Damasio's theory of the emergence of a moment of core consciousness from nonconscious cognitive-affective processes, resulting in a feeling of what is happening as our body-states are being changed through our interactions with our environment. Core consciousness is momentary and fleeting. Autobiographical (extended) consciousness adds the temporality of human awareness—our sense of being in and experiencing the flow of events in time, from the past to the present, and projecting into the future.

However, as we saw, the self that you are is not just your conscious self. Core and extended consciousness provide important dimensions of our self-identity, but they are far from the whole story, insofar as we are mostly carried along by nonconscious bodily habits of perceiving, valuing, thinking, and doing.

Where is the self in relation to these types of consciousness? Damasio's answer is that the self, as intrinsically embodied, exists already prior to and beneath our conscious awareness, but also includes all of our conscious processes, and incorporates aspects of mind (as a system of meanings) that extend beyond, and give meaning to, consciousness (Damasio 1999, 2010; Mead 1934). What Damasio calls the *protoself* is all of the nonconscious body and brain structures and processes that make up a relatively stable bodily organism, with its fundamental homeostatic and allostatic values. The *core self* roots in moments of core consciousness, which is the feeling of what is happening in the present moment as our bodily states are being altered by our engagement with our surroundings. The *autobiographical (extended) self* is the result of extended consciousness that creates, via memory of prior experiences, a narrative of our temporally developing identity. Let us consider each dimension in more detail.

1. *The Protoself.* The most fundamental task of an organism must be the maintenance of life within the permeable boundary of its body. In more complex animals, the homeostatic and allostatic processes are monitored by the brain's ability to map changes in the body-state resulting from its ongoing interaction with its changing environment (Schulkin & Sterling 2019). All of this life-sustaining activity has to be realized in the brain and body of the organism. The protoself is thus "*an integrated collection of separate neural patterns that map, moment by moment, the most stable aspects of the organism's physical structure*" (Damasio 2010, 190; italics in original). The protoself is what situates all our activities, physical and mental alike, in a body functioning within an environment. Most of these mapping processes are nonconscious, but occasionally we get conscious glimpses of their workings as felt changes in our body-states.

Damasio sees the protoself's ongoing management of, and response to, changing states of the body/brain as involving three different kinds of neural maps (Damasio 2010; Sterling & Laughlin 2015).

The first kind is *interoceptive maps* that provide information about our bodily states in relation to conditions for the animal's well-being (Swanson 2003b). These maps of our visceral well-being indicate how things are going for us with respect to our welfare (Denton 2005). Damasio introduces the notion of "primordial feelings" to capture the feeling of being alive, conscious, and functioning in a particular body as it engages its surroundings (Damasio 2010, 191). Interoceptive maps give rise to subtle feelings on a scale running from pleasure to pain, effort to ease, energy to fatigue, and so on. You could say that primordial feelings are what it feels like to be *you (or me), here and now, in this present living moment.* They play an important role in mood states, such as lethargy, invigoration, anxiety, desire, hesitancy, and so forth, which we feel in the deepest bodily way, as the source of our being in the world (Heidegger 1927/1962).

Damasio suggests that our sense of a unified stable self comes primarily from the relative stability of the homeostatic and allostatic processes, rather than from the body as a whole. The body is constantly changing and engaging in different activities, whereas "the internal milieu and many visceral parameters associated with it provide the most invariant aspects of the organism, at any age, across a lifetime, not because they do not change but because their operations require that their condition vary only within an extremely narrow range" (Damasio 2010, 194). In other words, if you are looking for that relative stability that you feel *as you,* then you might want to look at your most visceral processes of body-state monitoring, accessed via primordial feelings.

In earlier work, Damasio (1999) states explicitly that the protoself is preconscious, since it represents the body's nonconscious monitoring (via neural maps) of its states. That said, Damasio (2010), like James (1890), seems to grant that this basic sensibility can sometimes rise into consciousness, thereby generating our feeling of being alive and aware of our current body-state relative to our well-being.

The second kind, *master organism maps,* provide an overall schema of the physical body (with trunk, arms, hands, elbows, fingers, feet, knees, and head, etc.), and then body-part relations and motions can be mapped against the master map, thereby capturing the location, orientation, and movement of various parts of the body (i.e., the process of proprioception or kinesthesia).

The third kind, *maps of the externally directed sensory portals,* give us our sense of our various perceptual modalities as tied to specific body parts involved in those types of perception. We have a felt awareness that we see with our eyes, smell with our nose, touch with our skin (covering multiple body parts), hear with our ears, and taste with our mouth, tongue, and nose. Sensory portal maps thus appear to supply us with a sense of the perspectives from which we perceptually engage aspects of our environment.

2. *The Core Self.* The core self is based in meaningful activities of core consciousness. The move from the mostly nonconscious operations and feelings of the protoself to the emergence of a moment of conscious experience requires the protoself's mapping of external objects (a first-order map), its mapping of internal bodily states (a first-order map), and its mapping of how the external objects give rise to changes in our bodily states (a second-order map). Consciousness is thus a second-order mapping that provides us with awareness of changing states of our body as it engages its surroundings. A moment of consciousness is therefore "the feeling of what happens," as it is happening, for the one who is aware of what's happening. That feel for what is happening is not just about physical objects impacting our sensory systems, but also the feeling of our ongoing transactions with others.

In order for the core self to arise, there must be what Damasio calls a "non-verbal narrative" of the events experienced by the self: "Anytime the organism encounters an object, any object, the protoself is changed by the encounter" (Damasio 2010, 202). Because the protoself is being changed via its engagement with objects, the core self "is created by linking the modified protoself to the object that caused the modification, an object that has now been hallmarked by feeling and enhanced by attention" (Damasio 2010, 203). The result is a nonverbal narrative *of me*, as the protagonist in my own unfolding drama (and comedy of errors). It is this emergence of the protagonist that gives us a sense of a self that has "ownership" of its feelings and thoughts.

It is important to keep in mind that the "core" self does not require some inner homunculus or ego that is supposedly the substrate or locus of the momentary activity of core consciousness. Postulating such an ego simply moves the problem of the self to another level, that of finding some inner "thing" (a quasi mini-me) that is conscious by virtue of *having* certain feelings. To the contrary, it is the *whole person* who is conscious, by virtue of feeling changes in their body-state. As we saw earlier, in our treatment of Gerald Edelman's account, the condition for consciousness is massive reentrant mapping across diverse functional regions of the brain, within a narrow temporal window. The result is an emergent awareness of primordial feelings and other feelings that constitute our feeling of what is happening.

3. *The Autobiographical (Extended) Self.* Unless we have some brain malformation or suffer a serious lesion in specific parts of our brain, our consciousness is not limited to a razor's edge moment of core consciousness. Instead, it extends back into our remembered past and is part of an ever-developing future (James 1890; Schacter 1996, 2019). Extended consciousness (across time) makes possible a gradual, somewhat consistent buildup of memories of our personal past that forms the autobiographical self (Donald 2001).

   We will not attempt an account of memory here but will only note three important characteristics of memory that are especially relevant to the autobiographical self. First, memories are not stored whole in some dedicated "memories" closet in the brain. There is no such space. Instead, they are distributed across various functional regions of the brain involved in perception, feeling, valuing, and action. Memories (e.g., one's first kiss) cannot be plucked, whole and complete, from some inner memory storage facility. They have to be reconstructed, via association areas capable of combining these various sensory–motor images and their associated feelings and emotions. Second, memories are not simply in the head (Donald 2001; Clark 1998). They incorporate aspects of the worlds we are living in (Schacter 2019). Third, memories have to be consolidated, via certain stages of sleep, or else they cannot be recalled as a unified narrative flow of images (Tucker 2007).

Damasio identifies two mechanisms necessary for the construction of the autobiographical self: "The first is subsidiary to the core self mechanism and guarantees that each biographical set of memories is treated as an object and made conscious in a core self pulse. The second accomplishes a brain-wide operation of coordination that includes the following steps: (1) certain contents are evoked from memory and displayed as images; (2) the images are allowed to interact in an orderly manner with another system elsewhere in the brain, namely, the protoself; and (3) the results of the interactions are held coherently during a certain window of time" (Damasio 2010, 212–213). The result is the shaping of the autobiographical self via pulses of autobiographical consciousness.

Damasio's use of the term "object" for a memory needs to be taken cautiously. Memories are obviously not physical "things" in the way baseballs, bicycles, and buildings are things, but memories *are* unified images that can change the state of the protoself. Memories are *had* as complex images that affect the state of the protoself and ultimately the core and autobiographical dimensions of selfhood. Damasio's point is that, just as a moment of core consciousness (part of the core self) has as its "objects" things perceivable and capable of being acted upon, so extended consciousness (part of the autobiographical self) needs some "object" (here, a memory image or series of images) to activate a conscious state. The idea is that you are conscious just insofar as there is an "object" to be conscious of. As noted in chapter 5, this is what philosophers and psychologists call intentionality, the directedness of consciousness toward some object, which Franz Brentano took to be the mark of mind—the ability to intend an object (either physical or mental) (Brentano 1874/1973).

Damasio's key claim is that extended consciousness is sustained in pulses of awareness of the ongoing flow of images, as they affect the

> protoself. The images are realized by neural maps of objects and events, either physical objects and events, or memories (Donald 2001). The awareness (as feeling) of changes in the body-states of the protoself through its engagement with objects in its environment *is* consciousness. Extended consciousness requires an "image space" (Damasio 1999, 219) in which various types of images can be associated into a developing narrative of the autobiographical self, and it requires a "dispositional space" where memories can be reconstructed, feeling responses arise, and movements can be generated.
>
> The neuroscience of this is highly complex and also speculative. Damasio is well known for research on Convergence–Divergence regions of the brain where various types of sensory and motor inputs can be associated and organized (Damasio 1999, 2010; see also Griffin et al. 2015; Swanson 2003b). Briefly, the *image space* depends on the various sensory–motor areas in the brain that create neural maps and the images realized via those maps. The *dispositional space* depends on Convergence–Divergence regions, which coordinate inputs from other functional brain regions (e.g., early sensory cortices, higher order cortices, parts of the limbic system, and numerous subcortical nuclei) (Johnson & Tucker 2021). These Convergence–Divergence regions involve extensive reentrant mapping, consisting of mutual recursive signaling across functional neuronal assemblies which permits them to coordinate activations occurring in several regions within some narrow time window, thus providing a conscious memory (Edelman 2004).

Damasio's account of the relations among the protoself, core self, and autobiographical self (as based on nonconscious processes, core consciousness, and extended consciousness) helps to explain four characteristics typically attributed to the self: (1) you have a distinctive *perspective* on the objects being mapped by your brain that appear to you as mental images; (2) you experience a sense of *ownership* of feelings and thoughts that you regard as "mine," as inhering in *your* individual body-mind alone; (3) you have a sense of *agency* in relation to objects (including your body) that you can manipulate and move; and (4) you experience *primordial feelings* of what it feels like to be alive in your own body, at this present moment in time. Damasio seeks to explain these four characteristics, which seem to apply more to our sense of an individual mind with private experiences, while also emphasizing our social, communal nature and the intersubjective roots of our selfhood (see also Mead 1934; Schutz 1932/1967). In many respects, then, he is like Dewey in the way he addresses the intimate relation and interaction of mind, body, brain, and social landscape that conjointly constitute our bodily sensibility (Gallagher 2005; Johnson 1987; Schulkin 2004).

Damasio's explanation of these four basic features of selfhood runs as follows. *Perspective* arises from the externally directed sensory portal maps, which help us orient ourselves in our surroundings. The sense of *ownership* of "our" thoughts and feelings comes from the protagonizing of core consciousness, where the nonverbal narrative of what's happening is sensed as *mine* and rooted in *my* body. The sense of *agency* arises as we experience the "I can" (Merleau-Ponty 1962) of our bodies, which involves our awareness of body-part relations and feelings of motion in the body, directed toward objects in the world. The experience of *primordial feelings* is the basis for my capacity to feel what it is like to be me, here, now, alive, and feeling changes in my body-state. We could also add a fifth characteristic, which is the sense of *unity* of our selfhood, insofar as the self processes emerge from the relatively stable state of the homeostatic and allostatic processes of the protoself. Our sense of being a unified self is based on the stable patterns of the protoself, and not on the existence of a substantial, fixed self that, as Hume recognized, is nowhere to be found. This relative unity and stability of the self is fragile and can be dissolved by physical and psychological traumas that lead to various psychopathologies, such as dissociative disorders and schizophrenia.

The Narrative Context of the Self

Damasio's use of terms like "protagonist," "nonverbal narrative," and "autobiographical" self, suggests the narrative character of our selfhood. Dewey liked to emphasize the "dramatic" character of experience and self-identity. The self is an interpenetration of habits, many of which are shaped by individual and cultural narratives in the formation of our identities. Alasdair MacIntyre observes "Man is in his actions and practice; as well as in his fictions, essentially a story-telling animal. . . . I can only answer the question 'What am I to do?' if I can answer the prior question 'Of what story or stories do I find myself a part?' We enter human society, that is, with one or more imputed characters—roles into which we have been drafted—and we have to learn what they are in order to be able to understand how others respond to us and how our responses to them are apt to be construed" (MacIntyre 1981, 216).

This emphasis on the shaping power of narratives captures two crucial characteristics of our selfhood. The first is that human action is intrinsically narrative in structure (M. Gregory 2009; Johnson 1993; Ricoeur 1984), so that the self is formed up at least partly in and through its immersion in personal and cultural narratives. The second is that large-scale communally shared narratives are the primary ways we enact our cultural identities and our sense of values and purpose. Let us consider each of these key characteristics in turn.

Narrative structure is not an artificial framework externally imposed on an individual life. As Paul Ricoeur (1984) has pointed out, human action manifests an intrinsically narrative pattern that is rooted in the conceptual structure of action. Actions are not merely events connected via causal chains. The conceptual structure of action involves agents, character, acts performed, the historical context of the act, motives for acting, ends and goals pursued, and values to be realized. In prototypical narratives, protagonists strive for goals, struggle against obstacles, recruit allies, wrestle against antagonists, achieve successes, and suffer failures. So, from the get-go, actions manifest their dramatic story-like character. Ricoeur proposes that narrative is the highest level of synthetic activity for binding events together temporally into a meaningful story: "Time becomes human to the extent that it is articulated through a narrative mode, and narrative attains its full meaning when it becomes a condition of temporal existence" (Ricoeur 1984, 52). Our embeddedness in cultural narratives seems so natural to us because narrative is the form of understanding that best realizes the abovementioned characteristics of actions, and weaves them together into a temporal unity, helping us understand what happens to us. The various sciences provide other explanatory frameworks for events and actions, but they typically lack the narrative structure for understanding character, meaning, and action.

MacIntyre concludes that, "Narrative history of a certain kind turns out to be the basic and essential genre for the characterization of human action" (1981, 208). This leads to the second key factor, which is that the narratives are, for the most part, culturally instituted and preserved. Cultural narratives are not exclusively linguistic constructions. They are also realized in all forms of symbolic interaction—in gesture, ritual, institutional practices, painting, music, dance, sculpture, and other arts—that enact the meaning of our lives. Our culture's myths provide fundamental ways to understand where we came from, who we are, what is possible for us, how we ought to behave, and what we should value (Alexander 2013). In this way, our identity is influenced by the role(s) we are playing within narratives, including those into which we are born and those we consciously adopt.

Besides grand cultural narratives, there are our own more personal narratives that emerge from our individual development over our lifetime, such as being brought up in a certain family, at a certain time, in a specific place, under certain economic conditions, within a particular social stratum, and within communities of significant others. However, these more individual and personal narrative constraints are typically unavoidably intertwined with our larger cultural narratives.

The cultural narratives we inherit have a profound effect on our sense of who we are and who we might become. They are hard to shake. Even the testy adolescent who sees herself as rebelling against all of her culture's (especially her parents') smothering narratives nevertheless cannot escape narrative capture. She typically rebels by following the narrative role of *rebel*, adopting modes of reactive behavior that she shares with her peer group (e.g., mode of dress, hair style and color, piercings, offensive language, preferred anti-establishment music) At worst, we are captives to these narratives, and at best we are coauthors of them, only rarely crafting new narrative roles and identities.

We are suggesting that shared myths and life stories are one of the chief means of our culture reproducing itself and preserving a modest continuity of cultural development (Donald 2001). They constitute the primary way we off-load our values, conceptual systems, and institutions into cultural artifacts. From birth we are caught up in the narrative configuration of our lives through cultural story architectures that conform to a small, widely shared set of ways to discover and to act out our selfhood. The arts of diverse forms of self-realization constrain the range of options available to us, but at the same time they give us ways of molding our character via the roles we take up and personalize. The self-in-process is never an act of radically free self-creation. It is rather a cocreation, employing bodily habits, capacities, and values, and negotiated with significant others along culturally laid down paths of narrative development. It is an artistic experiment in self-formation, and, like all arts, it utilizes materials and means to create something new. The fact that our supposed "self-creation" is actually subject to numerous bodily and cultural constraints is not a bad thing, because exciting creativity is possible even under substantial constraints. There may only be a couple of basic sonnet forms, but consider some of the amazing and highly creative individual sonnets that have been created within those genre constraints. Likewise, there may be only a small number of generic narrative forms, but some folks have been remarkably creative in their particular realization of those forms, inspiring us and showing us what we might aspire to.

Dewey's Self as Interpenetration of Habits

Dewey, of course, had none of the neuroscience of the sort available today. What Dewey did so well, however, was to give a phenomenologically rich account of experience at the level of habit formation, rather than at the neural level. Dewey basically argues that "character is the interpenetration of habits" (1922/1988, 29). We propose to broaden the scope of the above definition by replacing "character" with "self." Habits as features of the organization of action are the diverse building blocks of successful adaptation. They underlie all aspects of the developmental and regulatory components of the self.

Dewey's most extensive account of the role of habits in the making of a self is in *Human Nature and Conduct* (1922). His focus there is on how our moral selfhood gets formed, and how it is possible to reconstruct the self for increased viability and well-being. We suggest that what Dewey has to say about *moral* selfhood provides a profound account of selfhood in general.

Dewey's conception of the self is, in certain respects, fairly straightforward. We are organisms with evolved sensory, motor, and affective capacities that continually interact with the affordances of our structured environments, which are a mixture of precarious and stable dimensions. We are eruptions of what Dewey called blind impulses, which, when taken up into a meaningful pattern of organism–environment interaction, become habits. "Habit is energy organized in certain channels" (Dewey 1922/1988, 54), and those channels of action are what give an impulsive upsurge of energy its meaning. A certain aggressive impulse becomes a habit of angry response, and the context of that meaningful eruption determines the character of that anger (e.g., a jealous fuming, a murderous rage, a righteous indignation, or a patriotic defense of country). Recurring patterns of interaction lead to the development of habits of perception, feeling, thought, action, and valuing. Repetition of experiences, in the face of recurring stimuli, strengthens habitual responses. Habits thus exhibit a tendency to harden and rigidify under the influence of repeated exposure to repeated relatively stable conditions, within similar recurring environments. Much of the time, this entrenchment of habit is a relatively good thing, insofar as it functions well as long as environing conditions are relatively stable. Clusters of habits contribute to the ongoing formation of a sufficiently stable self. Based on past experiences, habits can project expectancies for future experience, and they lead us to act more or less reliably on those predictions, all of this with little or no reflective thought required.

However, even though sedimented habits can be useful in familiar situations, they can become quite dysfunctional when new, unanticipated conditions arise that are different from those that generated our prior habits. In the face of these novel conditions, successful functioning requires a certain plasticity of habit (McEwen 2007; McEwen et al. 2015), making it possible for us either to reconstruct old habits, or to form new, more useful habits better suited to the new conditions.

The obvious question is, how is such plasticity of habit, as necessary for growth of the self, possible? Since reinforced habits tend to become rigid, how can we ever loosen them up in response to novel conditions? The answer to this important question is based on four basic characteristics of habits that we have previously described. The first is that *habits are recurring patterns of functional interaction between an organism and its environment, based on impulses that get appropriated into meaningful patterns of activity*. Habits arise in the ongoing attempts of an organism (here, a human) to navigate and negotiate its surroundings. From a neural perspective, evolutionary processes establish the general functional architecture of our brains and bodies, and then the course of our individual development over our life span creates functional neural clusters that, when they "fire together" are bound together into functional organizations we call habits (Graybiel 1998; Tucker 2007; Tucker & Luu 2012). The recurrence of instigating stimuli strengthens the tendency of that neuronal cluster to be activated under similar conditions (Jeannerod 1997; Rolls 1996). Most of the time, this strengthening of neural connections operates automatically and beneath the level of our conscious awareness.

Second, *habits incorporate aspects of the environment.* Habits are not just internal dispositions of the self, because a habit cannot be specified without reference to environmental conditions. There is an internal dimension, of course, in that habits rely on capacities of our brain and body for perception, movement, feeling, and action; moreover, habits incorporate bodily impulses into meaningful channels of activity. However, none of this activity can be separated from the environment in which it operates. Dewey observes that you can't have the habit of breathing without the air that you breathe, nor can you have the habit of walking without the ground or surface on which you move: "Habits are like functions in many respects, and especially in requiring the cooperation of organism and environment. Breathing is an affair of the air as truly as of the lungs. . . . The natural operations like breathing and digesting, acquired ones like speech and honesty, are functions of the surroundings as truly as of a person" (Dewey 1922/1988, 15). Habits reach out into the environment, and they are formed up in direct response to the affordances for experience and action supplied by one's surroundings. These organism–environment relationships are endlessly porous, and this factor will become crucial as we explore how habits can be changed.

Third, *habits constitute the self.* As we explore aspects of our environment, we establish habits of perceiving, thinking, feeling, valuing, and doing, rooted in our deepest biological, interpersonal, and cultural values. Clusters of acquired habits build up our self-as-presently-formed: "All habits are demands for certain kinds of activity; and they constitute the self. In any intelligible sense of the word will, they *are* will" (Dewey 1922/1988, 21). Habits are not isolated dispositional states. They intertwine and mutually affect one another, constituting the self as currently formed.

Fourth, *habits are arts for the transformation of experience.* When they are reflectively formed, habits give us intelligent ways to engage our surroundings. They are technologies, or arts, for our functioning in the world: "They involve skill of sensory and motor organs, cunning or craft, and objective materials. They assimilate objective energies and eventuate in command of environment. They require order, discipline, and manifest technique" (Dewey 1922/1988, 16). Habits thus establish our aesthetic sensibilities, thereby guiding our transactions with our natural and social worlds.

As necessary as habits are, they are not always sufficiently adaptive to enable us to solve the problems we face. When the ruts of habit become too deep, they become rigid, resistant to reflective critique, difficult to acknowledge, devilishly hard to revise. When conditions change, some habits formed under other previous conditions may become ill-adaptive, dysfunctional, and even immoral. For example, habits for interacting with students acquired decades ago may not be nearly nuanced and sensitive enough to deal constructively with the diversity of student backgrounds and situations that we find in today's classroom, given our current sensitivities. In the mid-1960s, one of our spouses had a much loved and sought-after high school teacher who had the practice of rubber stamping the word "toad" in ink on the forehead of a student who wasn't answering questions or paying attention. Back then, almost nobody gave this a second thought. It was just Mr. X being playful. Wasn't Mr. X so funny and charming? In radical contrast, in today's climate, he would be hauled up for harassment or creating a hostile learning environment, and rightly so. Habits that at one time seemed so in sync with the values and backgrounds of an overwhelmingly white, upper-middle class, homogeneous group of students, now generate offense, misunderstanding, alienation, and emotional suffering. They most likely had this effect on at least some students even back then, though nobody would dare to speak up about their feelings. As we have gradually become more aware of the deep and pervasive prejudices built into our cultural institutions and practices, and as we better understand the diversity of backgrounds, experiences, social strata, and opportunities among our students, we are faced with the need to change our habits. No more forehead stamping, no matter what!

This example might seem trivial, but it has nearly all of the features that characterize some of our most profound moral and social problems. It focuses on the sedimented, mostly unchallenged habits that constitute the status quo. It raises fundamental questions about the socioeconomic and cultural prejudices of our institutions and practices. It reveals the relevance of historical context. It calls for difficult self-criticism. It requires the hard work necessary to revise offensive, immoral habits. It reminds us how blind we can be to our deepest values and attitudes, and how very difficult it can be to change those values, beliefs, and practices.

The point of this example is that, absent any perceived problem, there is no motivation for change of habits. Often, the aggrieved in such cases are justifiably hesitant to speak up. The first step in moral deliberation is coming to see a situation as problematic. Only then is there any impetus toward reconstructive and ameliorative action. That is why one of the supreme gifts of people we regard as moral prophets and reformers (e.g., Susan B. Anthony, Mahatma Gandhi, Martin Luther King, Jr.) is their ability to recognize profound problems (e.g., deep inequities, marginalization, cruelty, oppression) and to open the eyes of those who have not, cannot, or will not acknowledge their complicity in an oppressive situation.

Once a problem is recognized, what is needed is the capacity to *intelligently* restructure habits and to initiate new ones. This is at the heart of Dewey's philosophy. The challenge is how to break out of the grip of sedimented habits to create new ones more responsive to emerging conditions. Dewey addresses this issue in the context of his account of the pervasiveness of the need–search–satisfaction process in response to changed conditions. Need arises when habits that define one's current identity—habits that were formed up mostly in response to earlier conditions—meet newly arising conditions that make those entrenched habits ill-adaptive. Often, the problem is the result of conflicting, incompatible habits (an incompatibility of perceived goods), either within a person who has multiple, sometimes incompatible, values, or between one's own values and those of others. The situation becomes indeterminate and the agent feels frustrated, confused, and unclear about how to move forward. Unfortunately, too often such encounters with a problematic situation lead us back to the mistaken view that all we have to do is turn back to prior habits, as if they *really were adequate* for solving the problem, if only we could see how.

Sometimes, though, if we are open-minded and self-critical enough, the problematic situation may set us on a search for transformative action, through reconstruction of habit and the making of new habits. It is this openness that so much characterizes Dewey's organic, naturalistic perspective. Where some see habits as mere constraints, Dewey would envision those constraints more constructively as providing an understanding of the problem at hand, and thereby influencing the contours of possible ameliorative action.

Reconstruction of habit does not occur in a vacuum, and it can be extremely difficult. Dewey was cognizant of the plights of people, the limitations, the hardships, and the inequities that permeate the social space. These prejudices that infiltrate our institutions, our modes of engagement, and our nonacceptance of others are viruses that can be so hard to eradicate. They infect our habits, our orientations, our values. These viral prejudices reflect our moral failures. Neither of us believes in the doctrine of original sin, but social sin, such as our transgressions toward other races, genders, ethnicities, and so on, is as common as apple pie. Indeed, regions of the brain (e.g., the amygdala) are quick to get the gist of things, but can also create a bias and a vulnerability, such as when we are too quick to discriminate on the basis of skin color (Phelps et al. 2000). Moreover, these habituated processes are not usually conscious (Phelps 2006). Reflection requires a continuous activity to monitor our prejudices and vulnerabilities, the less than better parts of our nature. Intelligence requires us to be not only smart, but wise in our decisions and responses, toward realizing what Lincoln in his First Inaugural Address called "our better angels." Dewey saw those better parts as directed toward resolution of conflicting values, enlargement of the circle of who we count as relevant in our moral deliberations (Kitcher 2011), freeing up and cultivating ours and others' capacities for valuation, and developing compassion for others.

Neuroscience research has only revealed the depths and extent of such biases, the kind we are not proud of (Kahneman 2011; Sapolsky 2018). But our social transgressions are part of the larger historical landscape, the images and depictions of people perceived as different from the dominant group. The amygdala in young children is less likely to demonstrate these historical biases, but no one is innocent. These are social constructions. We are not born with diverse social sins, but we sure do learn them easily.

Group formation—who is and who is not part of the in-group—that is so fundamental for our survival, can also be a vulnerability and a prejudice. Neuropeptides like oxytocin, critical for social attachment, contact, and empathy operate toward those who are part of the in-group, but not so for those in the out-group (Carter 2014; Sapolsky 2018). The same neuropeptide that can bind folks together in the in-group can elicit aggression for those outside the group. The neurobiology sets the conditions for our better angels, but we need to do the work to establish the appropriate enduring habits. Dewey understood that habit formation during critical periods in development is crucial for shaping constructive judgment and response. Human history is an endless testimonial of our failures of moral habit formation, leading to transgressions toward each other big and small.

Habits are the building blocks in the organization of action. Some are basic, deeper, and more highly activated across the brain. These habits will be less subject to change, while others will be more flexible. Our ability to orchestrate expanded use of capabilities is a feature of the evolution of intelligence (Rozin 1976, 1998). Problem solving requires intelligence in performing critical assessment of our habits of engagement, transactions with others, and transformation of problematic habits, often in cooperation with others.

Dewey described this situation in terms of his distinction between valuing and valuation (Dewey 1939/1991). *Valuing* refers to the fact that we desire and care about certain things, events, or states of affairs, which we are likely to experience as good. A diverse set of neural regions are intimately involved in the valuing process, including the amygdala, cingulate, and prefrontal cortex (Adolphs 1999; Schulkin 2004). That we value a particular something is a biological, psychological, or sociological fact about us, about our self as presently formed. We automatically and unreflectively prefer something that, for various reasons (e.g., biological, interpersonal, or cultural), we just happen to perceive as desirable and good.

On the other hand, humans are capable of intelligent reflection on a problematic, conflicted situation, and can explore possible reconstructive courses of action. We have evolved brain architectures and experientially developed patterns of neural connectivity that make it possible for us to examine the meaning and implications of our values. This use of critical reflection to discover what the problem is, how our current values are in conflict, and which options available to us are more likely to resolve the problem at hand, is what Dewey called *valuation*. Instead of just being driven by preestablished likings (i.e., valuing), valuation introduces an additional reflective capacity so important for our assessment, coping, adjusting, and learning behaviors, as key to our physical and social viability. Reflection is not a standing back from nature, or a hovering above nature, in hopes of acquiring some grand comprehensive vision, and then evaluating what has been "seen"; rather, it is embedded in, and is transformative of, natural events. Reflective valuation is itself an action that changes experience, rather than merely contemplating it.

How can we possibly move beyond the powerful inertia of prior habits that defines our current selfhood? The answer is plasticity of habit, realized in the ability to entertain multiple options for transformative action in response to a perceived problem. Dewey defines moral deliberation (in service of problem solving) as:

> dramatic rehearsal (in imagination) of various competing possible lines of action. It starts from the blocking of efficient overt action, due to that conflict of prior habit and newly released impulse to which reference has been made. Then each habit, each impulse, involved in the temporary suspense of overt action takes its turn in being tried out. Deliberation is an experiment in finding out what the various lines of possible action are really like. It is an experiment in making various combinations of selected elements of habits and impulses, to see what the resultant action would be like if it were entered upon. (Dewey 1922/1988, 132–133)

Dewey's view of moral deliberation is often criticized for being unable to provide any universal rule or decision procedure for which goods or actions ought to be chosen. Dewey's response was that, in fact, there are no such rules that could establish any algorithm for moral deliberation, and what principles or imperatives we have could never account for the emergence of new, unanticipated conditions and problems (see Fesmire 2003; Flanagan 2007; Johnson 2014a; Kitcher 2011). The alternative to such a rule- or principle-driven approach is an account of the virtues of inquiry that best serve our need to imagine how a certain projected course of action would "objectively" and more comprehensively resolve the tensions arising from competing values. Dewey suggests that dramatic rehearsal of the sort just described is more like creating and appreciating a work of art than it is like technical rule following. Whatever rules or principles one might articulate are just communally approved summaries of considerations deemed important in prior deliberative acts. Those principles may capture significant factors in responding to prior conditions, but they are never universal or eternal principles, since they may need to be revised or reinterpreted in light of newly emerging conditions (Johnson 1993, 2014a).

It is in dramatic imaginative rehearsal that our aesthetic sensibilities come most prominently into play. As you consider available courses of reconstructive activity, you rely on felt changes in the situation when one course of action resolves the problem better and more expansively. As we will see, in the following chapter on art and aesthetic dimensions of experience, we need to be keenly attentive to changes in the pervasive quality of the situation under consideration, trying to discern which option brings greater resolution of tensions, harmonization of activities, and liberation of capacities relative to matters of well-being. Just as we judge works of art, not for their satisfaction of some abstract principle of artistic goodness, but by their ability to open up new possibilities for growth of meaning and values, so also we assess possibilities for transformative action by how they enrich our grasp of the meaning of our situation and the meanings that could be realized by different courses of action. Judging possible actions is no more an activity of application of rules than is judging a work of art, though both activities rest on values that are subject to critical reevaluation.

Flexible Habits and Growth of the Self

The big problem lurking within this account of the formation and growth of selfhood is how it is possible to transcend the stranglehold of sedimented habits and to develop new habits, which is to say, creating an enlarged new self. We all know how powerful our many fixed habits are. Just consider the numerous factors at play in trying to change what you eat, initiate new exercise routines, or stop your behaviors that others may find objectionable (as in "Could you please stop rocking in your chair?" or "Do you have to chew your fingernails?"). Dewey describes the issue of habit change: "Habit is an ability, an art, formed through past experience. But whether an ability is limited to repetition of past acts adapted to past conditions or is available for new emergencies depends wholly upon what kind of habit exists" (Dewey 1922/1988, 48). From a neuroscience perspective, if habits are rooted in functional clusters of neurons that fire together and in sequence, and if the strength of those neural network connections is the result of repetition of actions in response to recurring stimuli, then how can we ever break out of the inertia of prior habits? You know, from hard-earned experience, how you continually fall back on preestablished habits, even as their repetition fails to produce good or desired consequences.

Dewey's answer to this problem is *flexible habits*. We are not doomed to repeat behaviors rooted in fixed prior habits (which constitute our present self): "A flexible, sensitive habit grows more varied, more adaptable by practice and use. We do not as yet fully understand the physiological factors concerned in mechanical routine on one hand and artistic skill on the other, but we do know that the latter is just as much habit as is the former" (Dewey 1922/1988, 51–52). Dewey admits that in his day no satisfactory neuro-physiological account of habit plasticity was available, but perhaps research over the past century is beginning to yield some insight into change of habits.

Dewey obviously lacked the neuroscience we have today. He formulated his account, therefore, not so much in terms of the science of neural connectivity and growth, but rather in terms of impulses and habits. Impulses (as energy pulses) arise from motivational systems and can be directed at certain objects or states of affairs, such as an impulse toward nutrition, and more specifically currently directed toward that chocolate cake you are eyeing. Dopamine is an important neurotransmitter in this motivational process in the organization of action (Berridge 2006; Schulkin 2007).

Impulses taken up into patterns of response to aspects of the environment become habits, and those habits give meaning and form to our impulses. The body is a vast source of often conflicting tendencies and impulses, but only some get appropriated into stable habits. We come into the world prepared for generating diverse types of activity that aid learning. Habits are built from these orientations. It is not chaos from the start. Even neonates have some motor control (Kagan 1984, 2002). Sensory–motor organization develops in a constrained context of possible movements. The neurostructure is binding from the start and grows more so in the development of habits of action and control (Graybiel 1998; Premack 1990; Premack & Premack 1995). Neural innervation shapes how we gain control of movement, action, and intention. There are always impulses, both latent and operative, that do not get taken up into habits. Consequently, in developing habits responsive to certain novel states of affairs, there are always more tendencies at work in the situation than get realized in particular habits. In this learning process, children are exceedingly good at watching others and imitating them (Meltzoff & Moore 1977). No magic, just evolutionary neural design geared toward mimetic activity (Swanson 2003b). Imagination is a critical part of this process. As we imagine possible actions, there are tendencies at work that may lead to new possibilities, ways of turning previously suppressed or sublimated impulses into new habits, into directed action—probing, constructing, fixing, altering, adjusting. Dewey saw learning as just this process of drawing on hitherto unregulated impulses to generate new meanings and new activities.

"Flexible habit" will seem like an oxymoron to most people. Isn't the point of habits that they are learned structures for engaging one's surroundings meaningfully, structures that solidify and become rigid through repeated activities? How could a habit actually be flexible? The closest

Dewey gets to a neuroscience response to this stability/plasticity problem is his claim that components in one habit may also be part of other habits. Consequently, the key to plasticity is to cultivate *more and more diverse* habits, rather than fewer, so that hitherto unappropriated impulses can surge up to figure in newly arising habits. A plausible neural account of this process might go as follows: In a particular functional neuronal assembly a group of coactive neurons fire together as part of acts of sensing, moving, and feeling. However, individual neurons and even larger clusters of neurons can be activated in multiple neural clusters underlying different habits. When certain subclusters of neurons are activated, they may thus be part of multiple habits. This creates a measure of indeterminacy in a particular situation, and it allows new impulses to arise and become incorporated into new or transformed habits. This is but a sketch of a possible solution, a sketch that needs to be filled in by further neuroscience research, but it suggests how inhibition of a specific habit might set the stage for new impulses to strut their stuff.

The Self in Process

Dewey summarizes his take on selfhood as follows: "Existentially speaking, a human individual is distinctive opacity of bias and preference conjoined with plasticity and permeability of needs and likings. One trait tends to isolation, discreteness; the other trait to connection, continuity" (Dewey 1925/1981, 186). The fragile, finite self is a combination of the constraints of sedimented habit that form our character (i.e., "opacity of bias and preference"), but which are sometimes interrupted with moments of plasticity leading to reconstruction of old habits and emergence of new, sometimes more adaptive, habits.

We can blend Damasio and Dewey. Damasio describes the emergence of core and then, extended, consciousness, as the body acquires the ability to sense changes in its state, in response to its engagement with objects in its environment, and not just objects but whole temporally developing situations, entire contexts. The self arises in three stages (or levels). The protoself is the developing set of nonconscious processes for (1) mapping of external objects, (2) monitoring the organism's own internal bodily states, and (3) feeling and mapping changes in one's body-state resulting from the object affecting the body. This engenders core consciousness, which provides a moment of awareness of how those body-states are being altered by changes in the environment in which the organism lives and acts. The core self is the protagonist of the nonverbal narrative of self-organization. Extended consciousness brings in the larger temporal framework necessary to construct an autobiographical self with memories of its past experiences and states. It is the immersion in mind of the core and autobiographical selves that allows awareness of the meaning of past experience, the significance of present experience, and the possibilities for future meaning and growth. Protoself, core self, and autobiographical self are not three distinct selves, but rather intertwined functional dimensions of one complex process of self-formation and development over one's life span.

Without any reference to Dewey's writings, Damasio provides a more neuroscientifically sophisticated theory of consciousness and self-formation that is compatible with Dewey's more phenomenological account of habit and selfhood. To flesh out Dewey's theory, we would need to translate impulse, habit, and flexibility into neural terms. Both Damasio and Dewey, in the end, supply us with a nondualistic account of selfhood, in which the developing self-in-process is embodied, embedded, enactive, and extended. There is no substantial "thing" that is the self, but neither is the self nothing. A person's self is who they are (and might be). One's current self is the product of one's evolutionarily established body and brain, plus its epigenetic development within various environments over the course of its lifetime. That self is an interpenetration of habits. These habits are not fixed in concrete, but can, in the face of novel conditions, yield to new impulses and self-organizations, giving rise to new habits. So, one's self is something, but just not a fixed, closed system. Neither is it completely open, entirely malleable, and radically unstable. Rather, it is a self-in-process, stable as it goes, and yet plastic in its creation of new order, new meaning, and new possibilities for action when it is faced with changing conditions.

In all of this, what remains as most important is the body—the living, breathing, moving, valuing body. It is the mediation through one's body that ties together our unconscious dimensions (the protoself), our conscious experience of qualities (the core self), and our temporally extended development of our narrative self (the autobiographical self). The body is not a fixed and finished thing. It is always changing. But it is as close to a stable self (as the protoself) as we are likely to get. So goes the naturalistic theory of selfhood.

Every Self Is a Moral Self

You might wonder how we could be nearly finished with our chapter on selfhood without once mentioning morality. Our reply is that, in fact, everything in this chapter is about moral selfhood, though not in the way that traditional moral theories construe morality. Dewey understands morality as a matter of what kind of character or self is being composed and developed as we engage in solving problems, reflecting, making value judgments, and acting in and through nature. There is no self prior to or outside moral engagement, because the self is inextricably caught up in moral relations with others and continually making judgments of better and worse. It is in this sense that to be a self—any kind of self—is to be a moral self. Let's consider the argument for this.

It can be tempting to think of the "self" as either an object or some sort of agency, possessed of certain perceptual, cognitive, conative, and affective capacities, and who then encounters a world in which these native capacities are put to work on experience. On this view, such a pregiven self then supposedly chooses which domains of experience (e.g., economic, moral, political, aesthetic, technological) it will engage with, at what point in time, and under what environing conditions. So, sometimes the self is acting as a moral self, sometimes as an aesthetic self, sometimes a spiritual self, and so forth. In each distinct kind of situation, the self allegedly utilizes the distinctive capacities that are thought to be relative to the type of situation being addressed. According to this view, sometimes we are making moral judgments, other times we are making aesthetic judgments, other times we are making economic judgments, and so on, for all of the different types of experience. A conscious being, endowed with these capacities to make these different kinds of judgments, then supposedly chooses whether to make such judgments and to act within those various domains of experience. These abilities to evaluate and make judgments require that we have learned which situations are about ethical decisions, which are predominately aesthetic matters, and which are about theoretical knowledge, to name but a few factors. We also need to learn what standards and values apply for each of these distinctive types of experience. Finally, we need motives to act appropriately and meaningfully in each kind of situation.

Going at least as far back as Plato, this way of carving up discrete domains of experience and types of judgments leads to the classic question "Why be

Moral?" The problem stems from the false belief that, even if we know what any particular moral tradition demands and expects of us, we still need a sufficient motive to do what we are morally obligated to do. This perennial separation of motive, character, and act leads to the mistaken assumption that we are passive beings that need some motivational *oomph!* to get us to do our duty, or to obey the call of conscience. It also suggests that we are a formed self even independently of moral relations, which we may or may not choose to enter into or engage with.

Dewey criticizes this whole conception of judgment, motivation, and morality, noting the artificial pseudo questions that it engenders. If motives are separated from acts performed, then the perennial question arises concerning what, if any, authority any standards or values might have over us, especially when we may be strongly inclined to do otherwise. Dewey recognizes that this problem dissolves once we have an adequate view of selfhood and motivation. The answer to the misleading question "why be moral?" is this:

> The authority is that of life. Why employ language, cultivate literature, acquire and develop science, sustain industry, and submit to the refinements of art? To ask these questions is equivalent to asking: Why live? And the only answer is that if one is going to live one must live a life of which these things form the substance. The only question having sense which can be asked is *how* we are going to use and be used by these things, not whether we are going to use them. Reason, moral principles, cannot in any case be shoved behind these affairs, for reason and morality grow out of them. . . . In short, the choice is not between a moral authority outside custom and one within it. It is between adopting more or less intelligent and significant customs. (Dewey 1922/1988, 57–58)

Seen in this light, we are always immersed in these practices, endeavors, and evaluations that we call "moral" because they involve considerations of better and worse ways of valuing and acting. We don't choose these practices. They choose us. The question is not shall we engage such practices and values, but rather how can we do this more intelligently, more constructively, more sensitively, more expansively? Our whole process of self-formation is a profoundly moral affair, because our actions inevitably affect the world for better or worse.

A key corollary of this naturalistic perspective is that moral considerations, aesthetic values, economic factors, political concerns, scientific issues, and so forth are interwoven themes of our selfhood. They are pervasive and intertwined in the fabric of our identity. Consequently, *the self is inescapably a moral self* (and also an aesthetic, economic, technological, spiritual self, etc.). You cannot choose whether or not to be morally involved. You just are, whether you like it or not. But you may sometimes and to a certain degree be able to affect the character, quality, and extent of your moral involvement (Sabini & Silver 1982). Any decision about better and worse by way of activity is a moral choice, according to Dewey's broad conception of that term.

With these observations about the pervasiveness and unavoidability of moral considerations, it is now possible to go back through the structure of this present chapter, to see that every description of the constituents of the self implicates moral values and evaluative practices. First, remember that the self is an interpenetration of habits. Impulses taken up into habits gain meaning, and the more these habits are strengthened, the more they become deeply entrenched constituents of our being. The self as presently formed is thus the product of prior habits woven into the complex fabric of character and identity. With respect to moral considerations, habits establish the channels of our choice and activity. Conflict of habits and values gives rise to moral doubt, rooted in the indeterminacy of the situation. Habits formed up in the context of past conditions can come into conflict, and become ill-adaptive, as new conditions arise. Moral deliberation is thus a form of problem solving; in this context, it is a matter of deciding what is better and worse regarding our values and actions.

Imaginative dramatic rehearsal is the form of problem solving through which we can project the possible implications of pursuing certain ends, giving preference to some values over others, privileging certain virtues, and following certain principles. In every process of moral deliberation, the self is at stake. We are deciding what kind of self we want to realize. The prior self, composed of prior habits, gives way to the newly forming self of reconstructed habits and new habits. The satisfactoriness of a moral deliberation can only be revealed in light of subsequent consequences and meanings realized through the transformative workings of imaginative rehearsal.

We have little control over future happenings, but we can judge the tendencies of certain dispositions, values, and practices, and in that way manage our developing selfhood with a modest measure of intelligence, foresight, and control. The "moral" self is not one self among several others, because all of the dimensions of our selfhood interact and interweave. For example, as we will see in the next chapter, our moral character includes aesthetic dimensions and values, not as add-ons from without, but as integral to our moral involvements. The self is just the complex blending of all of these many habits with their plural dimensions, falling in and out of sync with our surroundings (including other people), and undergoing continual transformation and, when we are fortunate enough, rejuvenation and intelligent growth.

8 The Aesthetics of Life and Mind

Truth be told, most folks don't spend much time explicitly thinking about the nature of art and its value for their lives. The small percentage who do reflect on such matters tend to recognize that, besides sometimes being pleasurable and entertaining, the arts are often sources of meaning and life values. Moreover, although we don't often contemplate these matters, art is omnipresent in our daily lives. Each of us experiences art and actively creates art as part of the fabric of our day-to-day existence, even though we are often unaware of doing so.

This present chapter is the longest in the book. Its length is a measure of the importance, for Dewey, of aesthetic considerations in all human meaning, thought, judgment, and valuing. Perhaps more than any other philosopher who ever lived, Dewey doubled down on the proposition that art is pervasive in our mundane activities. He saw that aesthetics encompasses, not just the making and experiencing of art and natural beauty, but all of the qualitative dimensions of human experience and meaning. He also saw that what we experience, value, understand, and do depends ultimately on our aesthetic sensibilities and values. In short, he went "all in" on the thesis that all meaningful experience is aesthetic experience. He argues that, if you want to learn how meaning is possible for humans, if you want to understand experience, if you want to know where our values come from, if you want to appreciate how morality is fundamentally constituted, and if you want to find meaning in your life, you ought to start with the pervasiveness of aesthetics in everyday experience.

The Devaluation of Aesthetics

You might wonder, if aesthetics is supposedly so important to who we are and what we experience and value, then why has that alleged fact been mostly overlooked, marginalized, and undervalued by society in general and even within mainstream philosophy of art over the past century. The answer is basically the association of art and aesthetic experience with feelings and emotions, which are not recognized for their important role in meaning. When we were graduate students in philosophy in the 1970s many of our fellow students and many of our teachers were fairly dismissive of aesthetics, regarded by most philosophers as the lowliest of the "value fields," with ethics, social, and political philosophy higher up on the scale of cognitive significance. The basis for this denigration of aesthetics was the mistaken view that because aesthetics focuses on the qualitative, feeling-based, and emotional dimensions of experience, it is too subjective to rise to the level of cognitive meaning, and rational operations of thought.

We have elsewhere surveyed some of the major developments in Western aesthetic theory that led to what Gadamer (1975) called the Kantian "subjectivisation of the aesthetic" (see also Johnson 2007, 2018, chapter 1; Johnson & Schulkin 2020). Since our primary purpose here is to develop an account of the constitutive role of aesthetics in everyday meaningful experience—what we elsewhere have called "the common currency of aesthetic sensibility" (Johnson & Schulkin 2020)—we will provide only a brief summary of the tendency of many philosophers to treat aesthetic dimensions as subjective and noncognitive.

You might assume that, if any field of study appreciates the role of feelings and emotions in our lives, it would be aesthetics and the philosophy of art. However true that may be, it is also true that Western aesthetic theory since the enlightenment has mostly regarded any judgments based on feelings and emotions as noncognitive and lacking in conceptual meaning. David Hume (1711–1776) and Immanuel Kant (1724–1804) had a profound impact on subsequent aesthetic theory, when they both argued that judgments of taste (i.e., judgments of beauty and sublimity in nature and art) are based on our passions and feelings.

Hume (1757/1965) had argued that aesthetic judgments of beauty are based on unique human sentiments (i.e., subtle refined feelings), and he explained how those sentiments could be cultivated by experience in appreciating art and natural beauty. Judgments of taste were generalizations about which objects and events evoke these socially cultivated sentiments.

Kant also defined *aesthetic judgment* as subjective and based on feeling: "Hence a judgment of taste is not a cognitive judgment and so is not a logical judgment but an aesthetic one, by which we mean a judgment whose determining basis *cannot be other* than *subjective*" (1790/1987, section 1, 204/44; italics in original).

However, contra Hume, Kant wanted more from judgments of taste, namely, that although subjective, they nevertheless could claim universal validity: "The beautiful is what is presented without concepts as the object of a universal liking. . . . For if someone likes something and is conscious that he himself does so without any interest, then he cannot help judging that it must contain a basis for being liked [that holds] for everyone" (Kant 1790/1987, section 6, 211/54).

The problem for Kant was that, if you relegate aesthetic judgments to subjective factors like feelings, then how can you explain the possibility of their demand for universal agreement? He assumed that concepts could have universally shared meanings, whereas feelings, being bodily processes, were idiosyncratic and entirely subjective. Aesthetic judgments are considered by Kant to have no cognitive standing (no conceptual meaning), because they are not based on shared concepts. So, the big problem was to explain how someone's feeling state, though subjective and not based on any determinate concept, can nevertheless claim universal validity (which Kant interestingly called "subjective universal validity").

We will spare the reader the details of Kant's arcane and highly ingenious argument concerning how a subjective judgment apparently grounded on feelings can nevertheless claim universal validity (for those details see Johnson 1978, 2018). For our purposes, the importance of this debate about the status of aesthetic judgments is the assumption that feelings are subjective bodily processes, not "cognitive" ones, and, as such, they cannot constitute the conceptual meaning of any given experience. This perspective thereby denies any meaning to the aesthetic dimensions of experience—a legacy that in some aesthetic theories has carried down to the present day. In short, the crucial consequence is the denigration of feelings and qualitative dimensions of experience as supposedly not capable of achieving cognitive meaning. This posture might well leave one wondering what the value of art and aesthetic experience might be, beyond entertainment. It was that concern that troubled so much subsequent aesthetic theory.

The nineteenth-century Romantic reaction against Kant's formalism and his denigration of feelings (as subjective) was a renewed emphasis on natural beauty, sublimity, and art as profound sources of universally shared meanings and feelings capable of helping us overcome our alienation from nature and recover a sense of meaning for our lives. Tolstoy, the great moralizing romantic, for instance, saw art as the expression and communication of basic emotions via artistic media in order to bind humans together in a fellowship of shared emotions: "Art is a human activity consisting in this, that one man consciously by means of certain external signs, hands on to others feelings he has lived through, and that others are infected by these feelings and also experience them" (Tolstoy 1898/1995, 511). Tolstoy believed that the most exemplary art was religious art, and especially Christian art, because it supposedly communicates feelings of love and care for others, and these shared feelings bind us into a universal "brotherhood" of humanity. Not all Romantics shared Tolstoy's explicit moralistic agenda, though most of them saw various forms of art as vehicles of emotional communion with nature and our fellow humans.

Wordsworth and other poets proposed art (especially *their* preferred type of poetry) as the means for overcoming our alienation from nature and other people. However, in privileging feeling over rational thought, they still accepted the dualism that pitted reason against emotions and feelings. They simply inverted the primacy of reason over the emotions, by elevating feelings and emotions over what they regarded as the impoverished technological rationality of the industrial revolution. Rational thought was regarded as being abstract and disconnected from our deepest human emotions. In its mechanistic operations it dulls and deadens human spirit. By contrast, nature and art were believed to engender emotional rapture, and the alluring tug toward moral and spiritual renewal, with art as a vital source of meaning and value.

The dawn of the twentieth century heralded in a new regime, known in philosophy as logical empiricism, which focused primarily on validating the objectivity of scientific knowledge, while rejecting abstract metaphysical speculation. Not surprisingly, the logical empiricist tsunami threatened to sweep away aesthetics, which was taken to be totally, or at least primarily, concerned with emotional expressions, lacking any cognitive meaning. The result was the near death of aesthetics at the hands of Anglo-American analytic philosophy, with its exclusive focus on science, logic, meaning, and knowledge (Ayer 1936).

In a desperate attempt to attribute some exemplary value to art, even if it had no cognitive value, Clive Bell's (1913) version of Kantian formalism ensconced art in a transcendent realm of pure formal significance, supposedly transcending the world of our practical, bodily activities of daily life. In what might be regarded as the epitome of a disembodied view of art, Bell notoriously argued that, in order to grasp the pure, transcendent form of artistic objects (i.e., their "significant form"), one must suspend all practical and moral interests, so that the art object can shine forth in all its resplendent formal glory.

This formalism was bolstered by the notion of aesthetic disinterestedness (inherited from Kant), which claimed that to experience the artwork fully, the perceiver must disengage from any practical interest the object might have in their daily life. This entailed suspension of any practical use of the object and disengagement from any life emotions and feelings the object might evoke in us. Supposedly, only by suspending our practical interests, values, and emotions could we grasp the full formal being of the artwork (Bullough 1912). However, having suspended all practical interests, and banishing all life emotions, one might be left wondering why we would value art at all, if it has no direct relation to our day-to-day concerns and values. Bell unfortunately tried to secure what he took to be the timeless value of art by elevating it to an eternal transcendent realm, thereby removing it from any sources of value in our everyday lived experience within our natural and social worlds.

If this sounds ridiculous to you, you are not alone. In the thin air of Bell's realm of art-as-significant-form, there are no emotions of daily life, nor any practical meanings. All that is left is a pristine aesthetic object shining forth and transcending the pitiful concerns and worries of our dull daily lives. Those few, select, refined folks who can grasp the awe-inspiring significant form of a work of art, do so via the experience of a unique "aesthetic emotion" that is alleged to be completely divorced from day-to-day experiences, thoughts, and feelings.

Such developments were the nadir of aesthetic theory throughout the first half of the twentieth century. Dewey's appreciation of the aesthetics of everyday life was bound to seem out of step with the epistemic assumptions of logical empiricism. When philosophical aesthetic theory began to reemerge in the late 1950s and early 1960s, it took the form of an "analytic" philosophy of art that emphasized an almost exclusive concern with the nature of key aesthetic *concepts*, such as beauty, sublimity, taste, art, aesthetic judgment, disinterestedness, form, organic unity. The importance of art for everyday life was overlooked in a quest to find a proper home for aesthetics in the suburbs surrounding the prized neighborhood of cognitive meaning and knowledge. The three most dominant historical orientations toward aesthetics during this period were Aristotle's naturalistic account of art as *mimesis* (leading to accounts of creative *mimesis*) (Aristotle 1941), Hume's theory of aesthetic sentiments (leading to emotivism), and Kant's formalism (leading to theories of universally valid aesthetic judgment).

Instead of investigating the making and experiencing of art, the supreme concern was conceptual *analysis* geared toward describing the cognitive status of various types of judgments and their implications for a theory of art. What resulted was meta-aesthetic reflections on the nature of the concept *art*, or of related concepts. For example, Morris Weitz, in "The Role of Theory in Aesthetics" (1956), argued, in a Wittgensteinian fashion, that *art* is an "open concept," one that can never be definitively defined, since new artworks will always arise to challenge previous essential definitions. Aesthetic turned into meta-aesthetics—reflections on the nature and status of a concept associated with art and aesthetic experiences. For a while, this type of analysis pretty much put the kibosh on exploring the meaning of an artwork *for our lives*, at least within philosophical aesthetics.

From this meta-aesthetic perspective (i.e., examining the concept of art, rather than our experience of art), Dewey's naturalistic appreciation of shared human aesthetic sensibility in everyday life was replaced with reflection on the concept of art, the concept of beauty, the concept of the sublime, the concept of expression, the concept of imagination, and so on. When this kind of conceptual analysis began to exhaust its linguistic resources, philosophers turned to the role of societal and institutional influences on what counts as art. George Dickie's *institutional theory of art* defined art by whatever art-world institutional standards are imposed upon certain objects and experiences within a specific culture, and at a specific historical moment (Dickie 1974). Dickie basically claims that art is whatever the "artworld"—consisting of artists, critics, gallery owners, museum curators, art patrons, and the consuming public—declares it to be, within a particular cultural context. Arthur Danto (1964) had earlier heralded "the death of art" as it was traditionally conceived, leaving us with artworks that turned reflectively inward to comment on and to challenge, traditional definitions of art and standards of aesthetic excellence. The days when artworks were regarded as exemplary enactments of human meaning and value appeared, according to Danto, to be passing away. In their place, we got art-oriented art conceived as a reflection on the conditions of art-making, a highly abstract and intellectual meta-aesthetic inquiry into activities within the art world. Think here of minimalist art, color field paintings, Warhol's *Soup Cans*, op-art, "happenings," feminist art, Judy Chicago's *Dinner Party*, conceptual art, and on and on. Then think of aesthetic theory directed mostly to explaining how traditional conceptions of art are being challenged in these new art forms, as they urge us to ask what art-world conditions make them *art*. Concern with the nature and significance of the art object is replaced by concern with the social institutions (the art world) that decide what counts as art in any particular historical period or culture.

There were exceptions to these dominant analytic views. One of the more notable was Susanne Langer's account of art as the cultivation of bodily form rooted in movement, gesture, and bodily expression (1953, 1972). Langer proclaimed that "music is the tonal analog of emotive life" (1953, 27), insofar as contours of human feeling are enacted by the patterns of musical tempo, pitch, and rhythm. Music was seen to be rooted in, and enactive of, the patterns of our sentient experience, such as the ebb and flow, and buildup and release, of energy in our bodies, tied to feelings of floating, leaping, rushing, hesitating, falling, pausing, and stopping. She saw arts such as dance as arising from the pervasive movement of animal life, reaching consummatory excellence of form and function in the works of choreographers like Martha Graham. Langer, a student of Whitehead and Cassirer, understood the importance of aesthetics in human life. However, she still separated feeling from cognition. Moreover, like her mentor Ernst Cassirer (1929/1953), she saw the evolution of mind as ultimately issuing in symbolic representation and she therefore gave symbolic representation center stage in her writing on art. She continued to see feelings in the context of epistemic concerns leading to symbolic interactions. Since the term "meaning" had been co-opted by the logical empiricists as entirely cognitive/conceptual, Langer had to coin the term "symbolic import" for the kind of body-based meaning that makes it possible for the arts to move us emotionally and to give us meaning. Being a woman in a profoundly patriarchal atmosphere, and challenging some of the founding assumptions of this philosophical framework, Langer had a tough row to hoe, and was not appreciated in her time. To Langer's credit, her appreciation of animal life is center stage, along with the intricate problem solving that diverse animals display and the human feats of expression and communication that develop out of those bodily perceptions and actions. She especially understood how the arts speak to our need for meaning by engaging the qualities and patterns of our sentient and emotional experience.

Another naturalistic thinker with pragmatist leanings was the nominalist Nelson Goodman (1978). He became known for his account of the arts as "ways of world making," and he championed the importance of art as being on equal footing with the sciences in the making of different worlds. He understood the importance of continual prediction and testing as a feature of science and art alike. That said, Goodman always kept separate the "worlds" made and he rejected the notion of a "nature" in which these worlds could be anchored. One comes away feeling oneself in the presence of someone who appreciated the role of art in our lives, but who had no strong sense of bodily sensibility as the key to human meaning and communication.

The 1970s saw the emergence of the deconstructionist and postmodern rejection of any essentialist thinking that might valorize the artistic consummation of meaning and values (Derrida 1967/1976; Barthes 1968). The valid motivation of these orientations was to challenge essentialist interpretations of *anything* (e.g., science, painting, literature, music, architecture), to undermine totalizing narratives, and to resist oppressive strategies for maintaining power to speak and act. Deconstructionists were rightly skeptical of the idea that works of art can express *the* meaning of an event or action, or that there could be a single proper literal interpretation of any text or artwork. But too often the baby got thrown out with the bathwater. In the service of noble liberatory motives, deconstructionists went to the opposite extreme when they criticized and often rejected terms like "presence," "meaning," and "value." While it is legitimate to challenge totalizing narratives and interpretations of texts of any sort, and to show how definite meanings are often undermined or "deferred" (one of their favorite terms), it does not follow that we must cease to regard various arts as vital enactments of meanings and values. The idea that no one meaning or interpretation of experience could ever be final and definitive is an important insight, but it does not require the denial of meaning in art and life. On the contrary, it reveals the richness, depth, and inexhaustibility of meaning. All our interpretations and value systems are always historically situated, perspectival, and based on values that might one day be challenged by new conditions, but this does not entail a rejection of the fundamental human eros for meaning (Alexander 2013). Self-critical awareness of the partiality and limitations of any interpretation of the meaning and value of an artwork is epistemically and morally of the highest value, but we can maintain such vigilance and still appreciate and inquire into the meaning of artworks for our lives.

In good Hegelian fashion, then, aesthetics and art theory in the twentieth century swung from one extreme—the supposedly objective analysis of the concepts associated with the arts—to the opposite extreme—the rejection of the very idea of stable meaning. As a result of both of these extreme positions, Dewey's focal emphasis on experience and meaning took a beating. From the analytic perspective, experience seemed too subjective and idiosyncratic, while from the postmodern perspective, experience seemed too foundational and too likely to lead to mistaken claims of absolute knowledge and definitive interpretations—claims that cannot be sustained in light of the perspectival and ever-revisable character of meaning, thought, values, and actions.

To oversimplify a bit, but only a bit, what's missing in most of these accounts of aesthetic experience and the arts is *the body*—the living, breathing, moving, copulating, making, dancing, singing, painting, sculpting, emoting, valuing body. This ignoring of the body led to overly cognitive and linguistic accounts of meaning, which were praised by philosophers doing conceptual or linguistic philosophical analysis, and loathed by deconstructionists, who were more interested in what Derrida called the dehiscence—unravelling or coming apart—of meaning.

It should come as no surprise, then, that in twentieth-century analytic or linguistic philosophy aesthetics has played only a very minor role. There were a number of philosophers interested in aesthetics and the philosophy of art (Arnold Isenberg, John Fisher, Nelson Goodman, Arthur Danto, Monroe Beardsley, Peter Kivy, and many more), but most of them tended to compartmentalize their aesthetic theories from their other philosophical concerns. The result was the relative marginalization of aesthetic dimensions in mainstream accounts of mind, meaning, thought, language, knowledge, morality, and sometimes even art. One fateful consequence of this bias was a profound failure to appreciate the role of aesthetics in our ordinary, day-to-day exploration of our surroundings and our attempts to find meaning and values for living. This was part of the larger bias of divorcing epistemic concerns from existential plights and wants.

Fortunately, the situation began to change in the 1970s. A growing number of philosophers began to pay attention to the role of the body in the constitution of meaning and thought, not just in the arts, but in every aspect of experience, thought, and action. First, the biological, cognitive, and neurosciences began to get into the act, providing a scientific perspective on how we experience and communicate meaning through art (Coegnarts & Kravanja 2015; Fauconnier & Turner 2002; Ramachandran & Hubbard 2001; Tikka 2008; Zeki 1999). Second, there was a growing appreciation of the role of aesthetic dimensions in everyday life (Alexander 1987, 2013; Johnson 2007, 2018; Leddy 2012; Light & Smith 2005; Saito 2019; Shusterman 2000, 2008). Also, the revitalization of pragmatist thought brought the body back into play (Alexander 2013; Johnson 2007; Shusterman 2008). These developments pave the way for a possible return to a focus on the aesthetic dimensions of *experience*, in all its complexity and richness, in a way that leads us back to a reconsideration of Dewey.

The Aesthetics of Meaningful Experience

Dewey entered the fray early on in the twentieth-century debates about art, in 1925 with *Experience and Nature* and in 1934 with *Art as Experience*. He carried forward a Romantic strain that valued the importance of art for a morally sensitive and well-lived life. He argued for an aesthetically rich conception of experience, cautioning against reductivist strategies that focus only on selected characteristics like sensation, form, unity, expression, knowledge, and imitation. As we saw in chapter 2, he observes how the notion of *experience* has become fragmented, impoverished, intellectualized, and subjectivized. One result is that we then restrict the aesthetic to only one narrow domain of experience, thereby failing to see how aesthetic characteristics permeate the whole of experience and give rise to meaning. The antidote to this impoverished view of experience is to see that aesthetics concerns all of the elements in our experiencing and making of meaning, rather than being limited to a small, marginalized set of allegedly unique experiences and activities called "aesthetic" and "artistic" (see Johnson 2007, 2018). Dewey summarizes the central theme of *Experience and Nature*, that all meaningful experience is art:

> Thus the issue involved in experience as art in its pregnant sense and in art as processes and materials of nature continued by direction into achieved and enjoyed meanings, sums up in itself all the issues which have been previously considered. Thought, intelligence, science is the intentional direction of natural events to meanings capable of immediate possession and enjoyment; this direction—which is operative art—is itself a natural event in which nature otherwise partial and incomplete comes fully to itself. . . . The doings and sufferings that form experience are, in the degree in which experience is intelligent or charged with meanings, a union of the precarious, novel, irregular with the settled, assured and uniform—a union which also defines the artistic and the esthetic. (Dewey 1925/1981, 269)

Dewey is not claiming that everything that happens is art. There are so-called experiences which are inchoate, fragmented, and undeveloped, which, consequently, have little or no meaning. Daily life is full of such empty occurrences. However, whenever the materials and processes of experience come together in a meaningful way, we have what Dewey called "*an* experience" (1934a/1987), a happening that is meaningful to us, based on what and who we are (our past, present, and projected experiences) and on the nature of the physical and cultural environments we inhabit. So, when Dewey speaks of "experience as art," the experience he has in mind is not merely humdrum, disconnected, and insignificant events of everyday life, but rather those enactments and consummations of meaning that matter to us. It is in this sense that a work of art is *an* experience, insofar as it is an exemplary or consummatory work of meaning-making.

The second half of the passage just quoted focuses on art (and experience as art) as a blending and harmonizing of the stable, fixed, and uniform aspects of experience with the precarious, indeterminate, and unstable dimensions. A work of art is consummatory in that it both carries forward and reworks the meaning of prior experiences, but also imaginatively projects new meanings, thereby opening up new possibilities for experience. Art realizes the continuum between meanings accumulated and preserved and the new meanings that may arise with the emergence of new conditions, thereby revealing new possibilities for understanding and action. A work of art provides and enacts both a richness of prior and present meanings and the stimulation of newly engendered meanings.

The aesthetic movement from the fixed and stable to the novel and transformative is rooted in our animal motivation and behavior. If we only experienced what is past and accumulated, we would fall into rigid inflexible habits generated via reinforcement of prior experiences and activities. It is the indeterminate and novel that motivates us, driving us forward toward new meanings and possibilities for search and satisfaction. This motivational arc holds just as much for higher-level cognitive operations of thinking and reflective knowing as it does for basic motivations that shape biological need–search–satisfaction activities. It is this sensed revelation of new possibilities that drives us to grow and flourish at all levels of our functional organization. This puts the individual in the driver's seat, not as a spectator, but as a motivated agent acting in the world to confront problems, imagine solutions, and search out consummatory situations

Art is exemplary of the blending of the ordered and settled with the contingent and novel that defines the pattern of ordinary experience. Art, therefore, does not merely copy (imitate, *re*present) preexisting patterns and events; rather, it *presents* and *dynamically enacts* both conserved and newly emergent meanings. It is for this reason that Paul Ricoeur (1984), interpreting Aristotle's remarks on art as *mimesis*, speaks of art as a *creative mimesis*, generating new possibilities (new "worlds"), rather than merely copying previous events and meanings (Goodman 1978). If *mimesis* were just copying, it would simply reproduce earlier conditions, and would ignore the appetitive lure of future possibilities for living. It would rob us of the motivations to act in light of new conditions. In animals, and especially in humans, it is the motivational allure of various perceived affordances that drives us onward to discover new resources, new chances for growth, and new ways of enhancing our well-being.

Art, as an intensification and deepening of meaningful experience, provides both a sense of meanings achieved and also new meanings yet to be realized through action. We can engage in exploration in this way without destroying the relative unity of our selfhood, since meanings discovered in past experience offer a stable basis within which novel events can arise. The self can be stable, but not fixed or finished. Dewey speaks of the "magic of poetry" and of pregnant experience alike as "the revelation of meaning in the old effected by its presentation through the new" (1925/1981, 270).

Experience and the Growth of Meaning via the Means–Consequences Relation

Art modulates between the extremes of rigid routine and capricious impulse. "Routine exemplifies the uniformities and recurrences of nature, caprice expresses its inchoate initiations and deviations" (Dewey 1925/1981, 270). This means that art has to achieve a tentative dynamic balance between the stable and the new. With a Marxist appreciation of the ills of alienated labor, Dewey sees that meaningful experience ("*an* experience") only occurs when the means to various consequences are themselves meaningful and not just a bare causal path to some desired end. The means must be meaningful, as *both* consummatory of prior meaning *and* productive of further consequences. The bare *cause–effect* relation thus gets transformed, by the addition of meanings, into *means–consequences* relations. Dewey observes that with the theme of the stable and the new, we have thus "insensibly passed over into that of the relation of means and consequence, process and product, the instrumental and consummatory. Any activity that is simultaneously both, rather than in alternation and displacement, is art" (1925/1981, 271).

The distinction between mere cause–effect and means–consequences is ontologically and epistemologically important. In Dewey's metaphysics of emergent functional organization, it is a profound step from the physical and psychophysical plateaus, where events cause other events, up to the plateau of mind, where means are part of the growth of meaning: "Means are always at least causal conditions; but causal conditions are means only when they possess an added qualification; that, namely, of being freely used, because of perceived connection with chosen consequences" (1925/1981, 275). Only creatures with minds, and thus partaking of meanings, can experience a means as part of the developing consummation of action directed toward an end, now conceived as a consequence, a meaningful desired outcome of action.

Why care about such a distinction? The answer is that the capacity to experience the continuity and interrelation of means and consequences is one of the keys to intelligent transformative action. By virtue of the emergence of mind, humans move from the bare push and pull of natural forces to the intelligent initiation of meaningful and purposive action that realizes new meanings. It is the crucial difference between being merely subject to energies and forces (i.e., causes) mostly beyond our control, and having a modest amount of control over how things play out in our lives. Means are not just causally related to ends; means are meaningful and appreciated as part of a consummatory process that results in desired consequences. Dewey summarizes the significance of this process: "For all the intelligent activities of men, no matter whether expressed in science, fine arts, or social relationships, have for their task the conversion of causal bonds, relations of succession, into a connection of means-consequence, into meanings. When the task is achieved the result is art" (1925/1981, 277).

Instead of speaking of the *work of art* (as a thing, indicated by a noun), we should recognize the *working* of art in our lives (as an activity, indicated by a verb). The value of the work is not just some imagined end state in which a series of causal actions terminate, but rather the experiencing of the developing meaning of the process that leads to certain consequences. What is meaningful is the entire means–consequences process, and we are in it, part of it, not separate. Means can be meaningful and consummatory as they work toward some consequence.

This valorization of meaning does not overturn the fact that life is full of meaningless activities. One is reminded of the second stanza of the Statler Brothers song, *Flowers on the Wall*: "Counting flowers on the wall/ That don't bother me at all./ Playing solitaire till dawn with a deck of fifty-one./ Smoking cigarettes and watching Captain Kangaroo./ Now don't tell me I've nothing to do." Yes, life is full of routine, directionless, and insignificant activities. Many labors have little or no intrinsic meaning, such as the bare act of splitting a piece of wood with an axe. However, if you embed that meaningless act in a purposive context, it immediately begins to acquire meaning, as means to consequences. Say you are splitting the wood for your elderly neighbor, who will happily pay you for your labor. The relation between the wage earned and the activity which garnered the wage is not intrinsic to the act, but extrinsic, since any number of different activities (e.g., mowing grass, building fences, and felling trees) might earn the same wage. But, what if you aren't doing the work for the monetary gain, or for any money at all, but just in order to help your neighbor? Then there is plenty of additional meaning, as your act contributes to your relationship, gives him something he values, and gives you the satisfaction of helping another vulnerable human being. Moreover, the very doing of the act can also give satisfaction in the artful accomplishment of the task at hand. There is the satisfaction of splitting a pile of firewood with power, accuracy, and efficiency. There is consummatory satisfaction in mowing a lawn well, building strong handsome fences, and dropping trees safely and exactly where you want them. All of these activities can be meaningful and consummatory in their own right, independent of extrinsic ends. Splitting wood, mowing grass, building fences, and felling trees are arts. The most mundane acts can thus acquire meaning—as workings of art—when the means are both consummatory here and now, but also lead to meaningful consequences. The splitting of wood can serve the maintenance of fire in the hearth, around which a family cuddles for warmth, play, conversation, and delight in the flickering flames and glowing embers.

It should be noted that the consequences of actions are not always positive and constructive. You might be helping your neighbor (a positive act) by felling trees and splitting wood for their wood stove, but through your action you may also be denuding the land, contributing to air pollution, and decreasing oxygen in the atmosphere. The meaning of one's acts is not limited only to the agent's intended consequences, nor is that meaning fully within their control.

The pervasiveness of art in our daily experience is the result of our capacity to grasp the means–consequence relations available to us in a given situation. In the sea of random, inchoate, and meaningless events that populate our daily existence, there emerge islands of meaning whenever mere cause–effect relations are realized as means to consequences, and thus acquire meaning, both for our present experience and for events to come (i.e., consequences). Dewey describes this process of inquiry into meanings as achieving ideas that "unite wide and enduring scope with richness of distinctions. The latter sort of consciousness is more than a passing and superficial consummation or end: it takes up into itself meanings covering stretches of existence wrought into consistency. It marks the conclusion of long continued endeavor; of patient and indefatigable search and test. The idea is, in short, art and a work of art. As a work of art, it directly liberates subsequent action and makes it more fruitful in a creation of more meanings and more perceptions" (Dewey 1925/1981, 278).

Dewey thus recognizes the fundamental role of the arts in everyday life, as part of the pervasive making and experiencing of meaning. We have needs for all kinds of activities that serve our desires and purposes, "but the characteristic human need is for possession and appreciation of the meaning of things" (Dewey 1925/1981, 272). Thomas Alexander observes that "at the root of our lives we manifest a deep-seated drive to *exist with a sense of meaning and value*" (2013, 140). The aesthetic dimensions of experience are an important source of meaning in our lives, and the arts are exemplary enactments of the possibilities for meaning. "The pervasive, qualitative horizon of the aesthetic dimension of experience is always present with us. We seek the possibility of fulfillment; action is a pursuit in response to the lure of the world with all of its possibilities. The fine arts, Dewey argues, merely raise this dimension of experience to full consciousness" (Alexander 2013, 139–140). Art can supply this fundamental need for meaning just insofar as it constitutes an interrelation of means and consequences in which both are conjointly and interdependently meaningful.

Dewey's more Romantic strain is evident in his conviction that works of art continually supply new meanings and new transformations of experience. This sense of the richness and inexhaustibility of meaning is no mere wishful illusion, for we see over the centuries how many works of art continue to evoke and enact new meanings across time and space. Consider for example, how Shakespeare's plays, in spite of his many parochial values and prejudices (e.g., about gender, race, and religion), nevertheless speak meaningfully to successive historical periods. Each new generation seems to find something in Shakespeare that addresses their concerns, such as historicism, psychoanalysis, Marxist social theory, existentialism, gender studies, deconstruction, critical race theory. Or consider the timeless power of certain music. In memory of Count Basie's passing, Judy Collins composed a song about the way he "sang the music of our souls," suggesting his broad, or even universal, appeal to the human spirit. In short, "the 'eternal' quality of great art is its renewed instrumentality for further consummatory experience" (Dewey 1925/1981, 274).

Leonard Meyer (1973), student of Charles Morris at Chicago and noted musicologist and pragmatist, emphasized how profoundly music appropriates our anticipation of the expected and our search for resolution for the unexpected. He saw aesthetic musical sensibility as manifesting the appetitive phases and the consummatory delight characteristic of our quest for viability and flourishing (Meyer 1956/1973; Schulkin 2013).

Substitute for "music" in the previous paragraph almost any activity (e.g., sculpture, painting, building, ritual, scientific experiment, mathematical theory, logical proof, philosophical thinking) and you find the same phasic movement from need to search to satisfaction (or dissatisfaction) that defines the creation and refinement of musical forms and genres. All involve active exploration of one's environment via processes of inquiry and problem solving that characterize our explicitly "aesthetic" search for meaning and values, for growth and consummatory experience.

If art *is* wherever consummatory meaning *is*, then art is pervasive in human experience. This requires us to attend to the aesthetic characteristics not just of obvious arts like painting, music, poetry, literary fiction, dance, architecture, and sculpture, but also to ritual practice, health practices, gesture, sports, communication, building of things, crafting of objects, socializing, designing and executing social and political institutional practices, and on and on, through all corners of daily life. All of these mundane activities can acquire aesthetic significance insofar as they involve means and consequences "charged with an office in quickening apprehension, enlarging the horizon of vision, refining discrimination, creating standards of appreciation which are confirmed and deepened by further experiences" (Dewey 1925/1981, 274).

The Common Currency of Aesthetic Sensibility

And so we come to see aesthetic sensibility as the common currency of our meaning-making across human activities of daily living (Johnson & Schulkin 2020). The roots of this sensibility lie in our evolutionary past, and the cultivation of this aesthetic responsiveness continues over the course of our individual development within some culture.

Dewey so admired Darwin's seminal work that he wrote an important article on "The Influence of Darwinism on Philosophy" (1910). Coincidentally, Dewey was born in 1859, the year Darwin's classic *On the Origin of Species* appeared. The scientific world would never be the same. Darwin, a shy retiring person, hesitant to come forward with his revolutionary ideas, was spurred to publish his work when it appeared that he might get scooped by Alfred Wallace. Darwin might have been shy, and Dewey might have looked shy, but both were bold evolutionary thinkers.

Chauncey Wright, a central member of the Cambridge "Metaphysical Club" (which included people like Peirce, James, and Holmes), embraced quite directly the naturalization of epistemic quests that was entailed by the new theory of evolution. Wright was perhaps the first philosopher on this continent to follow the consequences of the naturalization of human existence within the story of animal survival, consciousness, and viability (Fisch 1947; Wiener 1949; Wright 1877).

Dewey (1910) captured the excitement for philosophy of these new evolutionary ideas. The revolution was afoot, leading to the rejection of fixed forms governing species and frozen niches of habitation. These were replaced by processes of adaptive change within changing environments.

Dewey would note "we moved from the permanent to the changing" and he observed how "changes serve and defeat concrete purposes" (1910, 15).

Dewey avoided reductionistic accounts of the naturalism emanating from Darwin and others (e.g., Wallace, Huxley), and he was not tempted by any of the crude survival-of-the-fittest doctrines (e.g., Herbert Spencer) circulating at that time. Dewey drew on Darwin's understanding of natural selection and animal problem solving. Humans were now seen as firmly embedded in the natural landscape and utilizing available resources as means to desired consequences. There is no abstraction of the animal from nature, just the realization that we are part of it, in it, and that our capabilities are not frozen, but evolve in relation to our changing environment. He understood the integration of biology into philosophy, giving rise to a naturalistic sense of what we are about, what we can do, and what resources we can harness toward achieving some measure of well-being.

Darwin would write another great book, this one on the emotions. Here the diverse emotions that are generated mostly by mammals were explored in nonpejorative terms. Both Darwin (1872/1998) and Dewey (1896) tied the emotions to the adaptive landscape and resources. Emotions were not seen as something less than rational, but rather as a key part of problem solving. Emotions are neither passive, nor something that blinds our rational capabilities, but rather, they are an essential part of the bodily sensibility that shapes our exploratory activities and our foraging for satisfaction (Damasio 1999, 2018; Schulkin 2004).

Of course, not everything about the emotions is adaptive. Immobilizing anxiety, for example, is mostly not (LeDoux 1996; Rosen & Schulkin 1998). Nevertheless, fear is a signal of potential danger, mobilizing adaptive capabilities geared toward our survival and well-being. Facial expressions of diverse sorts, along with certain other expressive bodily behaviors, elicit approach responses, soothing responses, and feelings of comfort, while others elicit anxiety, fear, and withdrawal. Emotions for Darwin and Dewey were now seen as an equal partner with other capabilities that are more computational and have traditionally been understood as rational. The radical separation of emotion from thought had to be revised, in favor of an appreciation of the role of emotions in assessing the meaning of a situation and mobilizing intelligent adaptive problem-solving operations. The myth of pure dispassionate reason was replaced with diverse forms of problem solving rooted in motivational systems and shaped by emotions.

Moreover, it now becomes clear that our problem-solving capabilities and diverse ecological situated heuristics (Gigerenzer 2000) are rooted in our aesthetic sensibilities. Even in our daily chores and activities there are aesthetic judgments being made, though they are typically not conscious acts. The consummatory allure of affordances in our surroundings keeps us focused, interested, and motivated to persevere in our doing and making (Johnson & Schulkin 2020). This satisfaction necessary to carry us forward resides in the appetitive search processes that move toward resolution of problems and consummation of meaning. These processes operate in all phases of inquiry, just as much at the level of reflective thought and problem solving as they do in our maintenance of biological viability.

Animal survival requires continual assessment of both threats to well-being and identification of potential resources for life maintenance and satisfaction. Humans share these needs with other animals, but, given their enhanced capacities for grasping the meaning of things and events, they also appropriate these need–search–satisfaction processes for the highest levels of problem solving and the quest for meaning. In this way, our aesthetic sensibilities are appropriated for higher-level cognitive activities.

What, then, is the biological basis of aesthetics? The first thing, as Dewey well understood, is that aesthetics is not something outside the reasoning process, but something inherent in the organization of information processing, assessment, building, and constructing. There is no separate region of the brain supposedly dedicated to aesthetic sensibility. The reason for this is simple: there is no such center because our aesthetic sensibilities permeate virtually every aspect of our functioning. Aesthetics is inherent in the neurobiology of appraisal, search, and satisfaction or dissatisfaction. There is no unique dedicated circuitry in the brain for aesthetics (Schulkin 2013; Tucker 2007).

To be clear, there *are* dedicated circuits for many things, pain being one. However, scientists have abandoned the search for focal centers for pleasure, in favor of distributed systems active in the brain when approaching and avoiding something, liking something or not, being attracted or repelled (Berridge 2004). It is for this reason that, in looking for the sources of our aesthetic responsiveness and evaluations, we have to look at the whole person engaging her environment, which is to say, we need to look at the brain (in all of its distributed complexity), functioning in and running the body, which is in ongoing interaction with an environment of affordances.

Aesthetic processes cannot be located in any one part of this arc of experience, independent of the other parts. Likewise, imagination, that elusive process typically regarded as the key to artistic and scientific creativity, cannot be reduced to specific circuits in the brain. Imaginative capabilities are embedded in the diverse neural circuitries that underlie problem solving.

The nonexistence of brain centers or unique dedicated neural circuits responsible for our aesthetic sensibilities certainly does not make aesthetics less real or less important, because those sensibilities operate all of the time in our construction of objects, our movements, our singing, our dancing, and all our doings, physical and intellectual alike. It is just that these aesthetic processes are first located in the very mundane search for viability, and then they are recruited or carried up into our "higher" cognition.

One of the most important points we are trying to make in this chapter is that Dewey was right when he rejected the notion of distinctive types of experiences that require distinctive relations of various faculties, issuing in distinctive types of judgments. Since there is no uniquely aesthetic experience (to be contrasted with other types of experience), and therefore no uniquely aesthetic faculty to generate it, there is no point in looking around in the brain for a unique center of aesthetic experience. To repeat, contemporary neuroscience suggests that there is no such center. However, investigating neural networks, the operation of hormones, the basis for emotions, and reward systems in the brain and body is quite valid and important. We cannot understand the pervasiveness of our aesthetic sensibility without the contribution of biology, physiology, cognitive science, and affective neuroscience (Panksepp 1998). Aesthetic processes are just as robust a part of us as anything can be. There is no modular aesthetic faculty, but rather a multiplicity of functions distributed across multitudes of neural clusters, operating in parallel and via reentrant processes, and involving diverse hormones that affect our perception and action. So, one moral is to look for aesthetics first in the basic processes of survival and life maintenance, and then in surviving better, or flourishing.

The child comes into their world quite prepared to investigate and problem solve. Their cognitive–affective toolbox includes their ability to acquire language, general forms of problem solving, capacities for social interaction, an orientation toward natural objects, and a capacity to build, invent, explore (Mithen 1996). More specifically, there are special learning systems, such as the one that links what we eat with its impact on visceral distress or relief

(Garcia et al. 1974; Rozin 1976, 1998). This, of course, matters profoundly for a species like us, since distaste or disgust figures prominently in getting a sense of something, figuring out whether it is harmful or beneficial (Rozin 1976). Another example is the way we come prepared to discriminate many kinds of objects in the natural world, particularly those that are living versus those that are not (Mithen 1996). Also, quite early on, motion and some sort of agency are immediately linked in the neonate (Kagan 1984). Then there are all of the diverse constraints on information processing that enhance our evolutionary fitness. We thus come into the world fairly ready to learn, but radically underdeveloped compared to most species (e.g., the kangaroo).

At a fairly general level, the core of our aesthetic sensibilities is the recurring process of need–search–satisfaction. Early on, infants explore their environments and learn, in the most intimate visceral way, what satisfies and what repels. As Maxine Sheets-Johnstone observes, babies are born into the world in a state of "primal animateness" (1999, 137). Babies are squirming, wiggling, sampling, sucking, pooping, moving bundles of energy. At first, these impulsive bodily movements are somewhat random and undirected but, gradually, the infant's uncontrolled movements morph into meaningful patterned responses to aspects of their surroundings. In other words, they are learning the affordances of their world by turning impulsive projection into a more controlled search process that reveals sources of satisfaction and that enhances foraging activity. Infants begin to learn which means are embedded in, and lead to, which consummatory moments (i.e., consequences), both harmful and useful. Coupled with capacities of the sort described just above, the child develops their basic aesthetic sensibilities—their way of being in and acting in their world.

The consummatory affordances learned in our early development and extending over the course of our lives, shape our aesthetic responsiveness. For instance, young children spontaneously sing and dance. Their dancing, which often looks more like impulsive movement, is a natural extension of their primal animateness. As they dance to music, they are learning to coordinate sound contours with gestures and bodily movements, manifested in moments of joyful exuberance, fearful anxiety, and other states of bodily adaptation. As children learn to speak and sing, they pick up patterns of expression and communication that acquire new and expanding meaning. They are not honing some universal grammar module, but learning via their bodily interactions the meanings of various semantic–syntactic–pragmatic patterns. They learn, experientially, meaningful grammatical constructions rooted in the patterns of their interactions with their physical, interpersonal, and cultural environments (Feldman 2006; Lakoff & Narayanan in press). The whole shebang is a wealth of aesthetic performances and appreciations.

All of this adaptive growth and behavior involves emotions and feelings that have neurochemical and bodily bases. An adequate account of what goes into our aesthetic sensibilities would thus need to explain, not just the neural basis, not just muscular contractions, but also the key role of chemical signaling systems (e.g., neurotransmitters, neuropeptides, steroid hormones). An example of this kind of explanation would be an account of the role of dopamine in mobilizing the organization of action. Dopamine is essential for supporting and controlling the appetitive phase, such as in movement, statistical inference (i.e., prediction and expectation), (Schultz 2002, 2007; Schultz et al. 2008), responsiveness to salience (Berridge 2004), and to what matters for the organism relative to the affordances of its environment (Schulkin 2007).

In the appetitive phases of the search process, dopamine circuitry is activated from the brainstem to the forebrain, whether the event is about food, sex, or great visual art, music, or dance. Dopamine is neutral to our activities, meaning that it is active in anything we might be doing, any appetitive phase. Dopamine is just one neurotransmitter among others (e.g., norepinephrine, serotonin, GABA, glutamate), but it is an important one in the context of how appetitive urges are mobilized in the brain, generating behavior relative to our wanting something (Berridge 2004, 2006). This muscular neurotransmitter is built to aid our appetitive urges to understand and to engage our surroundings, the situations we are part of, and that afford us various needed resources.

Objects of aesthetic interest and fulfillment emerged, one would think, quite early in our evolution and expanded as we expanded our capabilities. Endogenous peptides, and chemical signaling systems in general, such as the endorphins, reach back a half billion years (Larhammar et al. 2009; Strand 1999). The diverse endorphins that are produced in the brain ameliorate discomfort, enhance the approach of a desired object, turn something less pleasing into something more pleasing (e.g., concentrated aversive tastes). They are important for hedonic consummatory experiences. In other words, a primary role for the diverse endorphins is the "liking" of events, the satisfactions and pleasures (Berridge 2004). Endorphins play a role, not just in pleasure, but also in managing discomfort or pain (Koob & LeMoal 2002). The rudiments of these events are ancient, mostly tied to basic needs. Then, endorphins (and other chemical signaling systems) extend beyond basic everyday viability to figure importantly across the expansion of human action and creativity, particularly in the arts, sciences, and sports. These modulators of pleasure and comfort eventually come into play also in epistemic capabilities of appetitive search and problem solving, easing a bit the unsettled state and increasing our sense of satisfaction and consummation that results from resolution of the problematic—the transformation of the unsettled into the somewhat settled or the very settled.

Aesthetic Sensibilities in the Experience of Art

We have been using the term "aesthetic sensibilities" as a catchphrase for a broad range of structures and processes in our bodies and brains as we engage our environments. The term "sensibilities" emphasizes our responsiveness to objects and events achieved by our bodily perceptual and motor operations. The term "aesthetic" emphasizes the qualitative, feeling, and emotional dimensions of experience and meaning. These dimensions are blended and bound together as part of our ongoing animal need–search–satisfaction activities, both at the level of our biological functioning and well-being, and at what we think of as "higher" levels of cognition and action.

Our aesthetic sensibilities operate across experiences ranging from the trivial and mundane up to high points of fulfillment and meaning. In other words, aesthetic sensibility is the common currency of our ordinary existence, playing its profound role in shaping what we perceive, experience, think, value, and do. This currency is at once both common and profound. It is common in that it shapes our basic biological and social functioning, and, as such, is a capacity we share with other people (and some nonhuman animals), but it is profound in being the basis for all our forms of meaning-making. That is what Dewey meant when he spoke of experience as art and art as experience in its exemplary manifestations.

A couple of examples from painting may help to illustrate what it means to say that artworks activate our aesthetic sensibilities and open up new meaning. How does a painting make a world? Consider David Hockney's

Garden (2015). The first thing to observe is that this is not world-making in the sense of *mimesis* as mere copying, but rather as creation of new significance—a *creative mimesis* (Ricoeur 1984). It is not a mere representation, not a *re*presenting of an allegedly pregiven fixed reality, as if the painting should or could be judged by its faithfulness to the original scene. Instead, it is a presenting of possibilities for meaning—of how the world can be for and with you. It is about the meaning of things and events. The painting enacts a world one is invited to inhabit through feeling and thought, a world to dwell in palpably, viscerally, and emotionally. *Garden* (2017) immerses us in a living, breathing, world-as-organism, into which we are inserted, not as spectator but as part of, and participant in, that world. You are not looking *at it*, so much as you are *in* and *of* it. Nature is alive, an effusive upsurge of rich organic diversity that yet somehow coexists in harmony. Nature is here plentiful and rich in meaning and values. The house, or bungalow, seems not so much dropped into nature as arising up almost naturally within it—not an intruder, but a welcome guest. The red walkway invites you, pulls you, guides you into an encounter with this effusive, comforting world.

Figure 8.1
Garden, David Hockney, 2015. Acrylic on canvas, 48 × 72". © David Hockney. Photo Credit: Richard Schmidt.

The colors capture the visual, eye-popping pizzazz that surrounds us in nature. The goal is not to copy or reproduce colors exactly (whatever that might mean), but to reveal the color contrasts, harmonies, and vibrancies in an immediate qualitative fashion. Also, notice how the "visual" is not merely visual. Your visual experience is inextricably interwoven with a sense of how things feel to the touch, are textured, and provide various degrees of solidity, manipulation, and movement. All of this is felt across multiple perceptual modalities, a testament to the multimodal character of perception. You are in this world visually, but the visual is interwoven with auditory, tactile, gustatory, and somatosensory awareness. With all of this teeming, pulsing, vibrant activity, the overarching mood is, nevertheless, one of an invigorating yet somewhat calm harmonization of these many impulses, qualities, and patterns of feeling.

Life, however, is not always a garden, a bowl of cherries, a field of sunflowers, or a romantic sunset at close of day. Artworks are consummations of possible meanings afforded by various situations, but those situations are not always joyful, enlivening, and heartwarming. Life has its tragic dimensions, its dark and oppressive side, its violence and destruction. In such cases, artworks can still be revelatory of the meaning of events and possible ways of being in the world, possible affordances. Contrast, for instance, Picasso's *Guernica* (1937) with Hockney's *Garden* (2017). This famous enactment of the horrors of war (here, aerial bombardment of the Basque town of Guernica in the Spanish Civil War) thrusts you into the midst of the chaos, terror, and anguish of war, not by thought or reflection, but rather by palpable immersion in the horror of that day. The world is black, gray, and white—no redemptive, healing, or enlivening color. *Everything* is torn asunder, smashed, ruptured, and destroyed. Limbs are severed and crushed. Bodies are ripped open. The ragged, jarring, disconnected forms capture the violent ripping apart, the dismemberment, and the demolition of life and home. The human and animal figures gesturally cry out in pain and terror, their anguish revealed in their screaming faces and wildly distorted bodies.

Guernica is indeed a profound act of world-making, but a horrifying world that could not be farther removed from Hockney's *Garden*. The worlds made by art are not always happy, but they can be consummations of the meaning of important events and objects, nonetheless. *Guernica*'s world is dark, massively violent, and viscerally terrifying. If *Guernica* provides a way to dwell in such a world, it is not by knowing where to hide from bombs,

Figure 8.2
Guernica, Pablo Picasso, 1937. © 2022 Estate of Pablo Picasso / Artists Rights Society (ARS), New York.

or how to fight back with bigger and more numerous bombs, but instead it beckons us to the seemingly impossible task of eradicating war from the world. It calls us to action, not by preaching the evils of war, but by physically, bodily, emotionally, qualitatively shaking us to our core and evincing feelings of repulsion at such evil, and such profound suffering.

Last, and least, we have to deal with the recurring question about Nazis. The objection sometimes raised against Dewey's view of art goes like this: If art is consummation of meanings, then doesn't that validate Nazi propaganda art, or any art motivated by hateful and vicious impulses, since that art is profoundly meaningful in the lives of hate groups and white supremacists? The answer is no. Yes, Nazi art and imperial architecture is art, and it no doubt supplied and still supplies certain nasty folks with meaning, with a world they might (and do want to) inhabit. But recognizing its power as art does not thereby make it good art, nor does it make it moral art. Art can operate for good or ill, so what we need to cultivate is our capacity for critical reflection about which values we ought to adopt and live by. Nazi art (as representing a certain kind of propagandistic art) leads to prejudice, exclusion, hatred, and cruelty. We have plenty of good reasons to think that such a world is bad. So, we have many reasons to criticize Nazi art as morally abhorrent and completely incompatible with the expansion, liberation, and enrichment of shared meaning and values. For these reasons, art alone does not get the last word. Critical valuation and moral deliberation get the last word, or should get it, if we want to make things better rather than worse.

Aesthetic sensibility is bound up with our embodied capabilities. When it operates primarily at the biological level, it seems to involve emotionally driven projections of expectancies for experience that are tested at the sensory–motor interface with our surroundings (Johnson & Tucker 2021). The search processes arise mostly from impulses and emotions that shape our behaviors. They seem, if not exactly mechanical, at least automatic and not under our conscious control. We are pushed and pulled by feelings and our emotional responses to perceived situations.

Humans, however, dwell in systems of shared meanings, so that processes that at the psychophysical level may appear to be dumbly rapturous impulsions, can become, in humans with minds, rich sources of meaning and understanding. The same feelings, emotions, and qualities experienced at the psychophysical level are still there, but now they acquire new meanings that can be drawn on to restructure our experience. We feel some measure of control, even if most of our actions still come from sources inaccessible by conscious reflection. What still remain basic are the bodily, affective, and qualitative resources that constitute our aesthetic responsiveness.

Our aesthetic sensibilities are thus the common currency of human experience, valuing, and action. The aesthetic dimensions of all the arts, and of experience as art, are intrinsic to our nature. Indeed, the cave art of fifty thousand years ago was an example of us wanting to control, to understand, and to make our peace with the objects in our world, both perceived and imagined. And what we see and imagine, what we hear and imagine, recruits the same neural circuitry. There is no extra circuitry for visual imagery or auditory imagery other than the neural circuitry designed for seeing or hearing, or for somatosensory exploration in an imagination space (Calvert et al. 1997; Kosslyn 1994).

Our aesthetic responsiveness permeates every corner of our experience, as we engage both our natural and created environments, other animal species, and human doing and making. Marvel for a moment at the communicative excellence of whale song, the shared motion of wolves and gazelles, the cheetah in fluid movement, the dolphins in their coordinated motion. Gaze anywhere and one finds the aesthetic dimensions of our participation in nature, in this case by tracking the wonders of natural excellence and beauty. Our bodily experiments in and with nature are shaped and flavored by the same aesthetics of inquiry that also operates in our more abstract conceptualization and reasoning.

Thought and Science as Art

Dewey always cautioned against rigid bifurcations, including the one separating science from art and aesthetics. All inquiry is a combination of aesthetic allures within the appetitive search mode, culminating in the consummatory mode, in satisfaction or delight. In his 1896 essay on the reflex arc, followed by *Experience and Nature* (1925), *The Quest for Certainty* (1929), and *Art as Experience* (1934), Dewey explored the common basis for art and science. He saw various sciences as great evolving creative arts, while he saw the arts as modes of exploration into the meaning of various experiences.

Dewey does not conflate science and art. Instead, he places them on a continuum of modes of discovering the meaning of experience. Science emphasizes the conceptual understanding of relations of objects and events, while the arts focus more on the qualitative dimensions of experience. However, even as their foci of attention might be different (i.e., science toward abstract relations and art toward qualitative richness), both science and art are modes of understanding of the meaning of experience.

On this view, aesthetic perception and artistic production are just different aspects of one process of the growth of meaning: "In esthetic perceptions an object interpenetrated with meaning is given; it may be taken for granted; it invites and awaits the act of appropriative enjoyment. In the esthetic object tendencies are sensed as brought to fruition; in it is embodied a means-consequence relationship, as the past work of his hands was surveyed by the Lord and pronounced good" (1925/1981, 280–281). So, aesthetic perception appreciates the richness of meaning worked up through past events and embodied in present experience of an object or event.

Artistic making focuses on enacting (i.e., exploring and realizing) possible meanings and future consequences, through creative organization of materials: "Artistic sense . . . grasps tendencies as possibilities; the invitation of these possibilities in perception is more urgent and compelling than that of the given already achieved. While the means-consequence relationship is directly sensed, felt, in both appreciation and artistic production, in the former the scale descends upon the side of the attained; in the latter there predominates the invitation of an existent consummation to bring into existence further perceptions" (Dewey 1925/1981, 281). Aesthetic perception and artistic creation are therefore revelations of the possibilities for meaningful experience that are rooted in our aesthetic sensibility.

Once art is seen to involve consummatory acts of meaning-making, its kinship with thought in general, and with science in particular, becomes more evident. We have been describing thinking as a manifestation of the basic need–search–satisfaction processes that characterize our ordinary biological functioning. If so, then the same kind of aesthetic perception and responsiveness that drives our organic biological functioning is also active in our higher-level cognitive operations of abstract conceptualization and reasoning. Once you see thinking as an embodied process that *makes* something, namely, *meanings*, it comes to be seen as art—the aesthetically guided exploration of meanings conserved and meanings newly discovered and enacted in various media. Dewey concludes, "Thinking is preeminently an art; knowledge and propositions which are the products of thinking, are works of art, as much so as statuary and symphonies. Every successive stage of thinking is a conclusion in which the meaning of what has produced it is condensed; and it is no sooner stated than it is a light radiating to other things—unless it be a fog which obscures them" (1925/1981, 283).

The idea of thinking as an art has radical implications for a naturalization of human mind, thought, and knowing. What we call antecedents and conclusions in a logical argument are just as much materials of thought as boards and nails are materials for the making of a house. What we call a conclusion is some object or event of thought whose relation to other events we are interested in. Contrary to popular opinion, in daily life we don't start with premises and work to a logical conclusion. Instead, we start with something we've taken an interest in, and then look for possible antecedent conditions that might have led to, and provided a justification for, that conclusion: "Premises are the analysis of a conclusion into its logically justifying grounds; there are no premises till there is a conclusion" (Dewey 1925/1981, 283). Dewey therefore thinks that we determine how good our selected premises are by seeing how well they seem to lead us to our selected conclusion, just as when we judge how good a certain paint is by how well it realizes a desired state of affairs for the siding of a house (e.g., protection from sun, rain, and rot, and resistance to fading). Deduction is concerned only with the relation of meanings to one another. It is not primarily about how those meanings do or do not stand in relation to experience, though the meanings involved will be rooted in experience. That is the sense in which deduction is a matter of logical validity rather than a matter of truth or falsehood.

There is no mind–body separation here. No divorce of self from world. No gap between thinking and doing or making. A good line of thought is one that reveals more new meanings and possibilities for experiences, and that helps us improve our engagement with our surroundings. All thought is a making and a doing that changes the world, rather than merely mirroring aspects of the world. It changes experience by instituting new meanings.

That brings us to acts of knowing—not knowledge (as a noun) but knowing (as a verb signifying an activity of making, organizing, remaking). The activity of knowing is the process of need–search–satisfaction whenever that process receives some measure of confirmation in experience. Since confirmation is always a matter of degree, a matter of more or less, it must always remain conditional and provisional—held as true according to current experimentation and evidence and subject to revision in light of changing conditions. Knowing is properly and more directly a result of inductive processes, rather than deductive relations: "The *act* of knowing . . . is always inductive. There is only one mode of thinking, the inductive, when thinking denotes anything that actually happens. The notion that there is another kind called deduction is another evidence of the prevalent tendency in philosophy to treat functions as antecedent operations, and to take essential meanings *of* existence as if they were a kind of Being" (Dewey 1925/1981, 285).

To say that induction is the only mode of thought involved in knowledge of existence is another way of saying that the need–search–satisfaction process is our primary activity for gaining reliable predictions about the possibilities of experience and their meaning for our lives. From this perspective, knowing is an activity of art, knowledge is a work of art, and science is just a more formalized and experimentally rigorous mode of art than we typically employ in our mundane acts of inquiry. The new objects made by the utilization of empirical methods are new meanings of what is experienced. Dewey summarizes this fundamental art of inquiry, leading to knowledge, and refined in science:

> Knowledge or science, as a work of art, like any other work of art, confers upon things traits and potentialities which did not *previously* belong to them. . . . Knowledge is not a distortion or perversion which confers upon *its* subject-matter

> traits which *do* not belong to it, but is an act which confers upon non-cognitive material traits which *did* not belong to it. It marks a change by which physical events exhibiting properties of mechanical energy, connected by relations of push and pull, hitting, rebounding, splitting and consolidating, realize characters, meanings and relations of meanings *hitherto* not possessed by them. (Dewey 1925/1981, 285; italics in original)

Knowing is thus an artful activity, creating new meanings and possibilities for experience through the working of the materials of nature. It is rooted in the common currency of our aesthetics sensibilities. It is empirical in its emphasis on testing of expectations (as hypotheses) against experienced objects and events. It is based on values, many of which come from our animal needs, capacities, and interests, and many of which come from our nature as social creatures dwelling in communities of inquirers.

What we call "science" and what we call "art," from a naturalistic perspective, are both art. Both are methods of inquiry into the meaning of things and events. Scientific method*s* (plural) may be more amenable to empirical testing and critique, but the arts can be just as rigorous, critical, and creative as any science. The sciences tend to focus on causal explanations and modeling, while the visual and musical arts tend to focus more on the qualitative dimensions of experience, but both are means for consummatory meanings. Consequently, to understand yourself, you most certainly should avail yourself of the best scientific research available from multiple sciences, but you also most certainly should avail yourself of what painting, music, poetry, dance, sculpture, and architecture can reveal about who you are and what your world is like.

Is there, then, only one method—empirical method—for gaining knowledge of our world? The answer is yes, and no. Method is empirical insofar as it involves an animal with needs for survival and well-being that must project expectancies for experiences and then test those predictions (as hypotheses) against what is actually encountered in experience. This is the only sense in which there is one empirical method. But, that said, there is no one way to be empirical, no one way to carry out the critical inquiry required. Insofar as there are at least three planes of existence (i.e., physical, psychophysical, and mind) and there are multiple levels of functional organization in the human brain and body, there will be a plurality of more specific modes of inquiry and explanation (e.g., evolution theory, physics, chemistry, systems and molecular biology, physiology, psychology, behavioral and social inquiry, and neuroscience). Dewey notes that "thinking is pre-eminently an art" (1925/1981, 292). Therefore, empirical thinking (as induction), and science as refined empirical method of inquiry, are just as much arts as are the manipulation of physical materials to create artworks that invite aesthetic perception.

Finally, we come to the value of art in human society. Dewey's answer is fairly straightforward, and it follows directly from what we have said in this chapter: The arts are exemplary enactments of possibilities for meaning within a situation. They are technologies of meaningful experience, in the Greek sense of *techne*, as the use of intelligence to help us deal with the contingencies of our existence (Hickman 2001; Nussbaum 1986). We unfortunately often distinguish the "fine" arts from the "practical" arts, which results in the badly mistaken view that the fine arts have no practical value in our lives. This is the opposite of the truth. Dewey concludes that "fine art *consciously* undertaken as such is peculiarly instrumental in quality. It is a device in experimentation carried on for the sake of education. . . . The creators of such works of art are entitled, when successful, to the gratitude that we give to inventors of microscopes and microphones; in the end, they open new objects to be observed and enjoyed" (1925/1981, 293). We would only add to this description of "objects to be observed and enjoyed" Dewey's central point that the arts are also to be utilized in service of our viability, our well-being, and our sense of meaning and value. Amidst the drudgery of disappointment and heartache in inquiry are the aesthetic delights that keep us going. No surprise that aesthetics is so common, so pervasive, and this is just the sort of thing that Dewey recognized so profoundly in *Art as Experience*.

9 Philosophy Naturalized

Philosophy Impoverished and Philosophy Revitalized

Candidates for the Catholic priesthood, along with their Protestant counterparts, are often required to declare their vocation, their calling to serve God. That's a pretty heavy burden of responsibility, thinking of oneself as called by God to do special things, like preaching, counseling, baptizing, and celebrating the Eucharist. Carpenters, electricians, and plumbers—folks who clearly do make our lives better—do not typically see what they do as a divine calling, but as both a way to make money and to exercise a useful skill that they find meaningful and satisfying.

Although philosophy students usually do not think of themselves as having a divine calling, they often do see themselves as having an important special intellectual mission—a rational calling, if you will. That mission is to help people become more reasonable, broader in critical perspective, and more reflective about themselves and their world.

This is a perfectly good educational mission, that of cultivating our capacities for more sensitive, expansive, and empathic intelligence for living. However, there are at least two major ways to carry out this mission, and they take us in two very different directions. One perspective sends us off in pursuit of transcendent sources and foundations of knowledge and values. The other sends us in search of meaning and values in the natural world. The former, transcendental, perspective is often based on the idea that there are "perennial" philosophical questions about matters of ultimate human importance, like the nature of mind, consciousness, selfhood, being, meaning, thought, knowledge, causation, logic, truth, and morality. The goal is to gain objective knowledge of these topics, via conceptual analysis and rational reflection.

When we (Jay and Mark) were in graduate school back in the 1970s, we and our fellow students quickly learned what the big philosophical questions were supposed to be, and we learned who the bigshots were—the thinkers who could supposedly show us how to engage those perennial issues. We were encouraged to apprentice ourselves to those thinkers, in order to learn their views and all of their analytic and critical moves and parries. The unspoken challenge was to learn to think like the "Big Boys" (masculine noun intended!), such as Plato, Aristotle, Descartes, Kant, Hume, Hegel, and Nietzsche, plus a host of more recent lesser stars, such as Frege, Russell, Wittgenstein, Carnap, Hempel, Popper, Quine, Putnam, Goodman, Davidson, and Rorty. The narrowness of scope and the uncritical exclusivity concerning *who* could be a member of these in-groups has now been amply demonstrated in scores of books and hundreds of articles over the last forty years. One of us had a professor, a well-known logician, who, only half-jokingly, proclaimed that there were only two great philosophers—Frege and Russell! He followed this pronouncement with something like "Yea, even the birds in the trees chirp their names: 'Frege! Frege! Russell! Russell!'" Philosophy at that time was predominantly a white man's club and, if a woman wanted to join, she had to demonstrate her brilliance and analytical rigor by taking on any of the approved list of respectable philosophical problems, and by engaging the approved philosopher stars of the day. One perspective for writing a history of Anglo-American philosophy of the last century, would be to focus on the progressive widening of the circle of *who* was allowed into the "house of philosophy" and also *which topics* were to be taken seriously (McKenna & Pratt 2015).

This view of philosophy engendered in some students a feeling of being part of a long and revered tradition of great thinkers who we hoped to emulate and whose brilliant insights we sought to channel. Our mission, should we choose to accept it, was to be the bearers of rationality in a fallen world. Even if it was not verbalized as such, we came to think of ourselves as bringing the intelligence, logical rigor, and analytic skills necessary to tackle these most profound human problems about mind, life, and world. So, we had our calling; not a divine one, but still a pretty exalted one—or so we thought. One not so salutary consequence was the tendency toward the arrogance of thinking that philosophers could analyze and evaluate the underlying assumptions and claims to knowledge operative in other disciplines, while implying that folks in those other fields didn't really have much to offer to philosophy. Moreover, under the pressure of the felt need to master concepts like "inscrutability of reference," "indeterminacy of translation," "paradigm shifts," and "value-ladenness of theory," many philosophers were slow to appreciate the emerging contributions of new voices that had previously been excluded or at least marginalized, but which now introduced new perspectives and new problems.

Dewey thought any such sanctification of philosophy was a mistake, insofar as it trafficked in erroneous and misleading notions like the transcendental ego, absolute knowledge, necessary truths, intuitive insight, universal reason, pure reason, and categorical imperatives. He was no fan of philosophy as something akin to a divine calling, although he saw philosophical reflection as a very important part of an intelligently lived life.

Dewey found the landscape of logical empiricism, the dominant emerging force in his day, to be existentially sparse, if not barren. The primary focus of logical empiricism was to explain the nature, conditions, and limitations of empirical (especially scientific) knowing, regarded as the epitome of objective knowledge. Notions like "sense data" and "observation sentences" were supposed to make it possible to empirically ground our knowing. Thought was taken to be conceptual, propositional, and linguistic. According to logical empiricism, concepts and the propositions by which they were organized into knowledge claims were taken to be cognitively meaningful by virtue of their grounding in sensory experience (i.e., the empiricist part), and then they were available to be combined and elaborated via a universal logic (i.e., the logical part). It was a good trick, if you could pull it off, but it turned out that nobody could, and for important reasons (see Hempel 1965; Kuhn 1962; Quine 1960; Wittgenstein 1953).

Dewey objected to this approach to philosophy for at least three reasons. First, he thought that reducing philosophy to epistemology (i.e., seeing philosophy as fundamentally about cognition and knowing) left out vast expanses of experience and meaning, thereby leaving philosophy mostly irrelevant to human existential concerns. Second, he thought their failure to address the key role of our aesthetic sensibilities in our experience and thought profoundly impoverished philosophy. Third, he rejected the idea of philosophy as seeking to accumulate truths about the world, which would then make philosophy a competitor with the scientific disciplines for discovering universal truths, thereby treating philosophy as science on steroids, the ultimate pursuit of objective knowledge and truth.

The logical empiricists, in return, pounced on Dewey. He was deemed not empirical enough. He was deemed not clear enough, because he tended to smudge all of their fine logical distinctions and dichotomies. Where were the formal logics (e.g., propositional logic and the predicate calculus) that were the darlings of the empiricists? Where were real experiments? When a student expressed an interest in Dewey, the question was "Why? Why would you bother to wade into those shallow muddied waters?" It is ironic that the logical empiricists thought Dewey had little of value to offer, especially when their conception of philosophy—as concerned mostly with analyzing concepts, solving logical puzzles, and revealing paradoxes—spoke little or not at all to the existential and moral concerns of ordinary people. Granted, logical empiricism did give us some insight into the logic and growth of scientific knowledge, but it had almost nothing to contribute to a philosophy of life and human values (McCumber 2001). Stanley Cavell's praise of the ordinary (1988) showed how certain recurring philosophical issues concerning skepticism, uncertainty, knowledge, and values do speak to our daily lives, but he was just one voice in the large choir of linguistic philosophy, and his explorations focused mostly on issues surrounding skepticism.

It is worth noting how different Dewey's philosophy naturalized is from Quine's epistemology naturalized (Quine 1969). They both eschewed anything transcendent, supernatural, and nonempirical (though they had very different views of what "empirical" means). They both rejected many traditional dualisms. However, Quine's focus was much narrower and much shallower than Dewey's. Quine was concerned mostly with the logical status of *beliefs* and their role in conceptual systems, theoretical explanations, and knowledge. In a behavioristic vein, it was all about sensory irritations leading to beliefs, which were not mysterious mental entities, but rather dispositions to specific behavioral responses. Although Dewey shared some of these sentiments, he regarded the exclusive focus on knowing as one of the chief sins of philosophy, leading us to ignore the many other important dimensions of experience that make us who we are. He saw knowing and all other forms of thinking as fully embodied, shaped by values, and rooted in aesthetic sensibilities. In short, Dewey's naturalism applies across all aspects of philosophy (e.g., metaphysics, epistemology, logic, ethics, political theory, aesthetics, spirituality), while Quine focused more narrowly on logic, epistemology, and ontology, and, within those fields, even more narrowly on structures of belief and their relation to sensory stimuli and observable behavior. The starkest contrast is thus between Dewey's focus on our aesthetic sensibilities as the key to meaning, understanding, and values, versus Quine's logical interests that neglected our aesthetic sensibilities.

Dewey offered a naturalistic conception rooted in the aesthetics of everyday life. Dewey's naturalistic perspective finds its own calling as the pursuit of wisdom—understanding the meaning of experience and bringing criticism to bear on our value systems, in order to determine what is better and worse by way of belief and action. This whole process is aesthetic at its core. It becomes more humble and more humane as it investigates value systems and their implications for our lives. It becomes conscientious concern for values to live by, not as prescriptions handed down by the gods, universal reason, or some transcendent source, but as rooted in our mundane attempts to make sense of our world and to make things better. That is the key—amid lots of conflict, confusion, and competing voices, to imagine better paths for the situation at hand.

Following Dewey, we have been emphasizing the human knower and doer as an embodied agent foraging for some measure of viability. However, knowing is just one part of who we are and what we do. We are biological, interpersonal, aesthetic, moral, cultural, political animals, and all of that has to be figured into our understanding of human nature. Though there is a place for prediction and control via propositions operating in a broad web of semantic meaning, as Quine believed, Dewey emphasized the whole person exploring their situation to make sense, to suggest possibilities, to render things that matter meaningful and open to enhancement and development. Both views (Dewey's and Quine's) emphasize the empirical nature of satisfactory methods of inquiry and the integration of science into the larger understanding of the knowing process, but Dewey's naturalism focuses, not just on acts of knowing, but on the whole human creature as immersed in and fully engaged with their physical, social, and cultural worlds. We are born into conflict and struggle. Our way forward is doubtful, fraught, and seldom clear. Philosophy naturalized gives us some critical tools for addressing our life problems as best we can, as we engage critically and seek rich and diverse meanings.

Dewey was sanguine about the possibilities for making things better. He recognized what matters most in our lives. He showed us that our own self-interest was not merely a matter of individual satisfaction but extends out toward the shared well-being of our larger community. He explored a path that opened up possibilities for humans, and there was nothing naive about his optimism, forged as it was in the course of his long life and the tragic events he suffered through, such as the death of two of his young sons. For him, intelligent inquiry expands our possibilities, our capabilities, and our freedom. Dewey was eminently sensible, not dogmatic. He stayed close to what mattered—human meaning, intelligent inquiry, worthy possibilities, normative goals, criticism of values, and enhancement of the human condition in the most comprehensive way.

Values and Criticism

Dewey's conception of philosophy focuses on critically assessing the meaning of our value systems. This undertaking is compromised and made more difficult by our tendency to radically separate natural events (facts) from values. He observes that, once you rip values from natural processes and events, they end up either as Platonic ideals or else as merely subjective preferences, and you will never, ever be able to appreciate how values are woven into the very fabric of our lives, from values built into our biological functioning, all the way up to our most important moral, aesthetic, and scientific values. Once values are separated from facts, the result is, on the one hand, a world of natural existences and, on the other, a completely different world of values (either absolute or completely subjective). Values are then seen as externally imposed on natural events. A more misguided dualism would be hard to find.

The only remedy for this radical divorce of facts and values is not to buy into it in the first place. Once you recognize that values are everywhere in nature (human and nonhuman), there ceases to be a problem of how values can attach themselves to states of affairs among natural events. Values are in, and of, natural events. We must overcome the temptation to think of values as merely subjective, as contrasted with natural events, which are supposedly objective. Many a pragmatist has made this point, following Dewey (e.g., Neville 1974; Putnam 2000; White 1947). Once values are seen to be part of natural goings-on, the proper question is not about where various values come from. We have plenty of excellent explanations of the natural sources of our many values. Instead, the important question concerns how conflict of values can be resolved and which of the many conflicting values vying for attention in a particular situation ought to be cultivated and enriched through inquiry and action.

Consider, for example, the many values embedded in farming practices, learned over centuries of ongoing investigation and criticism of existing practices, in order to improve them. These values have emerged from the *practice* of farming, by trying things out to see what works "best" (which can only mean *better*), and under which conditions. Farmers learn that certain fertilizers dramatically increase yields and the quality of fruits and grains, but their experiments may also make them aware of unwanted—even life-threatening—effects of some of those chemicals. They have to weigh the pros and cons of pesticides and fungicides, and their decisions can have substantial, very fateful, and extremely long-term consequences. So, there is a conflict of goods—increased yields versus potential damage through runoff in rivers, effects on wildlife, toxicity in humans, depletion of certain resources in the soil, and threats to the health of ecosystems—and farmers have to deliberate and choose what is to be preferred, and under what conditions. The values involved are not just ecological, but also economic, social, moral, medical, and aesthetic. Moreover, since these deliberations concern qualitative matters, there is no algorithm for deciding which value(s) ought to be preferred in a given problematic situation.

The significant point here is that the values are rooted in natural events and human practices, developed over decades, centuries, and millennia, and they can be criticized, reordered, and improved through inquiry and transformative activity in the natural world. There is nothing subjective about the values generated in such an evaluative process. It is a matter of experimental inquiry, open to criticism, tried out under many conditions, refined over time, and, yes, always fallible and subject to revision.

The key to understanding this process of value inquiry is the distinction discussed earlier between valuing and valuation. Recall that *valuing* is simply the state of preferring one thing over others at a given point in time: "Values are values, things immediately having certain intrinsic qualities. Of them as values there is accordingly nothing to be said; they are what they are. All that can be said of them concerns their generative conditions and the consequences to which they give rise" (Dewey 1925/1981, 297). *Valuation* is this latter kind of inquiry into means and consequences—an investigation of the generative conditions of specific values and their implications for future experience, in search of what is better and worse.

Valuation thus introduces the factor of critical assessment. Valuation is criticism and reflective judgment. In a situation marked by conflict of values, criticism examines the meaning of particular values by revealing their generating conditions and their consequences in experience. This is the only way we have of going beyond mere liking of something (i.e., valuing), in order to achieve an appropriate critical perspective (i.e., valuation). Dewey concludes that "every intelligent appreciation is also criticism, judgment, of the thing having immediate value" (1925/1981, 298).

Criticism of this sort is what we mean by being rational, or reasonable. It is an evaluative process that is close to home, personal, and respectful of the experience of others. It is a matter of acknowledging, broadening, and enriching experience. Emphasizing experience is a way to keep us honest, and maybe somewhat more responsive to others and their concerns and plights. Dewey reminds us that we are in this together (see also Scanlon 1998). He never lost track of that, as he went around the world, adjudicating the ills of the Trotsky trials, promoting some sense of what participatory democracy needs to be, both abroad and at home, and serving as a founding member of the National Association for the Advancement of Colored People. Like the rest of us, Dewey was no saint. He had his blind spots. Did he do enough on racial injustice? By present standards, probably not. Did he recognize the cultural richness of Native Americans? Probably not (Pratt 2002; McKenna & Pratt 2015). Did he fully grasp the dark underside of "progress?" Not completely, though he was more cognizant of this than most of us. Nonetheless, racial and ethnic equality, meaningful labor, intelligent sustainable progress, and maintenance of cultural traditions were key values for Dewey.

Philosophy as Criticism of Criticisms

Philosophy is a form of critical inquiry into the generative conditions and the consequences of values, relative to consideration of better and worse with respect to flourishing. It is thus a form of criticism, directed at whole systems of values and criticism: "Philosophy is inherently criticism, having its distinctive position among various modes of criticism in its generality; a criticism of criticisms, as it were. Criticism is discriminating judgment, careful appraisal, and judgment is appropriately termed criticism wherever the subject-matter of discrimination concerns goods or values" (Dewey 1925/1981, 298). Philosophy is a "criticism of criticisms" because it critically evaluates, not just individual values, but also the relative worth of competing systems of valuation/criticism. It seeks to find out what systems of criticism best serve our inquiries under certain specified conditions within a situation.

Philosophy thus strives for a certain generality, insofar as it aims for the big picture when it comes to our competing systems of value. Philosophy is pursuit of wisdom—an art of living—rooted in cultivated and nuanced criticism, or critical method(s). Philosophy should make no pretense of discovering absolute values. Rather, it is about cultivating the critical sensibility necessary for assessing competing systems of valuation.

Value conflict of this sort is a basic feature of the human condition, both in the internal milieu and in our social and cultural transactions. The goals in managing such conflict are "(1) to live, (2) to live well, (3) to live better" (Whitehead 1929b/1958, 8). To flourish, we need to look for avenues of life that allow both ourselves and others to enrich their values and free up their problem-solving capabilities. We know that short-term satisfaction is an endless allure, built into our appetitive nature. However, to live well, we must see beyond the present glittering moment, to embrace the longer-term possibilities for harmonizing, enriching, and liberating relevant values. Determining what these possibilities are is too often a random, rambling, blind process, but it can be made more reflective and intelligent with the aid of what Dewey called a "metaphysical ground-map of values," as the basis for a system of criticism (see below). Progress in criticism is not guaranteed in advance. It is a series of steps—big steps, little steps, frustrating steps, hard to see steps, steps backward, and no steps. There is no ultimate Google map route for how to get to our envisioned destination, our temporary end-in-view. We have to work out these possibilities with the aid of general critical reflection on competing systems of values.

Our greatest resource in this quest for well-being and well-doing is critical intelligence: "Because intelligence is critical method applied to goods of belief, appreciation and conduct, so as to construct freer and more secure goods, turning assent and assertion into free communication of shareable meanings; turning feeling into ordered and liberal sense, turning reaction into response, it is the reasonable object of our deepest faith and loyalty, the stay and support of all reasonable hopes" (Dewey 1925/1981, 325). This is the basis for what Dewey called our "common faith" (1934b/1986)—a faith in human intelligence as capable of helping us live more meaningful, more harmonious, more humane, and more creative lives. We should be humbled by how precarious this undertaking really is, and how fleeting the relatively stable goods are that matter most to us—trying to live well, managing endless worries, finding worthy goals, valuing others, keeping ourselves mindful of our human condition, and joining in the labor required to make things better.

The moral labor necessary for improving our own well-being and that of others is not denigrated by getting one's "hands dirty" with life's contingencies, by coming down from the heights of theory to immerse oneself in the hard labor of trying to resolve conflicting values. In the Aristotelian framework, manual labor, as involved in management of change, was viewed as less valuable than theory, which attempted to grasp the necessary, the fixed, and the complete (Arendt 1958). In contrast, Dewey actually elevated thought by grounding it in labor and problem solving, and by integrating it with practiced valuation. Theory is improved by the ongoing labor of testing and probing, subject to continuous feedback and criticism. This is the work of an investigator trying to make sense of things in a problematic context. Philosophy naturalized appreciates valuation as a fundamental achievement of philosophical acumen.

What, then, makes for better and worse with respect to modes of criticism? Dewey answers, "Criticism is reasonable and to the point, in the degree in which it extends and deepens these factors of intelligence found in immediate taste and enjoyment. Conscience in morals, taste in fine arts and conviction in beliefs pass insensibly into critical judgments; the latter pass also into a more and more generalized form of criticism called philosophy" (1925/1981, 300). There is a continuity, then, among the critical processes operating in aesthetic judgments, moral judgments, and philosophy-as-criticism-of-criticisms. All are forms of valuation, leading up to philosophy as the evaluation of systems of valuation. If there is any legitimate sense of philosophy as having some higher calling, it is based on this generality in criticizing various systems of criticism, or valuation (Neville 1974; Putnam 2000; Weissman 1993).

Philosophy thus becomes the practice of criticism directed toward the causal antecedents and causative consequences of specific value systems. Consider three examples of what this critical process might look like in ethics. First, since Kant has had such a profound influence on subsequent moral philosophy down to the present, we would want to interrogate Kant's conception of *respect* in order to determine its generative sources and its implications for future experience. Did Kant need to ground respect in what he called pure practical reason, or could its value be derived from, and realized in, experience? What is the basis for a certain type of creature to deserve respect? Second, in light of the impact of utilitarian theories, we would want to consider the viability of using a consequentialist moral calculus for computing the relative value of competing goods within a utilitarian moral framework. Is the character of our lived experience amenable to calculations of the amount of goodness of projected acts? Are such calculations even possible given our psychological makeup? Is there any role for calculations in moral deliberation? Third, with the recent upsurge in moral psychology, virtue theory has gained a new respectability. So, we would want to examine whether virtues can be rank ordered into a system capable of guaranteeing ethically correct behavior? Are virtues, as traits of character, even possible from a psychological and neuroscience perspective? These are but a few kinds of representative questions that would fall within the purview of philosophy as a general form of critical evaluation of value systems.

Dewey, as you would expect, argued that each of these moral systems (rationalist, utilitarian, virtue based) captures selected salient aspects of our experience of moral evaluation, but no one of these critical systems alone gives a suitably rich account of moral deliberation (Dewey 1930b/1988, 279–288). No single methodological or theoretical perspective can be hegemonic, so we have to assess the origins, limitations, and consequence of each system of values, interrogating the meaning, usefulness, and relevance of these three types of moral system under varying conditions (Rawls 1971). The development of moral judgment is geared in part to understanding how, under which conditions, and to what extent, these orientations are applicable to human problem solving and are supported by empirical research (e.g., Flanagan 1991; Johnson 2014a; Rawls 1971).

Concerning the nature of philosophy as criticism of criticisms, Dewey concludes, "Philosophy is and can be nothing but this critical operation and function become aware of itself and its implications, pursued deliberately and systematically. . . . These values, criticisms, and critical methods, it subjects to further criticism as comprehensive and consistent as possible" (Dewey 1925/1981, 302). Philosophy is one important tool toward self-reflection, development of intelligence, and attunement to what counts. Philosophical undertakings are a very natural part of inquiry, assessment, reflection, and discovery, all of which are rooted in our basic processes of inquiry, learning, and evaluation (Johnson & Tucker 2021). Philosophy demythologized—freed from the illusion of comprehensive and certain knowledge—is just a part of how we develop an evolving system of cultural values.

Dewey is often criticized for his emphasis on technology, and control in general, as cures for the human condition and its ills (Niebuhr 1932). However, Dewey proffered a rich and expansive notion of technology as the uses of human intelligence, or other tools, geared toward management of the contingencies of experience, via problem solving and reconstructive activity (Hickman 2019). Understood in this broad way, systems of inquiry and valuation can be seen as intellectual technologies, and philosophy can be seen as contributing to our cultivation of appropriate technologies of valuation. There are no generic one-size-fits-all technological cures for the plights of the human condition, but there are multiple intellectual technologies applicable to the kinds of problems we humans face. Some of these are concerned mostly with our individual capacities to control our behaviors, to inquire intelligently, to problem solve, and to imagine meaningful activities. Others are more socially oriented, involving ways to expand participation, help others live meaningful lives, and minimize our tendency to trample and harm others (Hinde 2002; Kitcher 2011; Rorty 1989). Dewey had his eye on the prize: acknowledging injustice, facing up to our diverse array of social sins and transgressions, but moving ahead, under the guidance of our cultivated philosophical imagination.

There has been a strong temptation in philosophy, and especially in moral philosophy, to separate facts from values, and then, when the question arises of why certain values apply to specific natural events, it is asserted that the values somehow carry within themselves their proper application to concrete situations. This view is mistaken. Dewey points out that "the conception that there are some objects or some properties which carry their own adequate credentials upon their face is the snare and delusion of the whole historic tradition regarding knowledge, infecting alike sensational and rational schools, objective realisms and introspective idealisms" (Dewey 1925/1981, 303). The mistake here is to regard as valuation what is merely a valuing. An act of valuing may mistakenly treat some object or state as an intrinsic good, when what is really needed is critical valuation, to determine the meaning and relevance of certain values in a situation of conflicting values: "When the question is raised as to the 'real' value of the object for belief, the appeal is to criticism, intelligence. And the court of appeal decides by the law of conditions and consequences. Inquiry duly pursued leads to the enstatement of an object which is directly accepted, good in belief, but an object whose character now depends upon the reflective operations whose conclusion it is" (Dewey 1925/1981, 303).

Cultivated and nuanced criticism is thus the antidote to habits of valuing that have become so deeply entrenched and strengthened that they appear to us as self-evident and in no need of revision in relation to newly arising conditions. Our selfhood gets formed in and through the reinforcement and rigidifying of such value habits. It is no easy matter even to criticize these values, let alone change them. "It is easier to wean a miser from his hoard, than a man from his deeper opinions. And the tragedy is that in so many cases the *causes* which lead to the thing in question being a value are not *reasons* for its being a good, while the fact that it is an immediate good tends to preclude that search for causes, that dispassionate judgment, which is prerequisite to the conversion of goods *de facto* into goods *de jure*" (Dewey 1925/1981, 303).

Immediately following this passage, Dewey makes a somewhat surprising claim about a possible candidate for a supreme value, when he has hitherto seemingly been arguing against any notion of an absolute or supreme value. You might have expected him to say that no value is good in itself (intrinsically good), because the preferred value in any given situation has to be determined only by ongoing critical inquiry within that situation. What Dewey in fact says is: "Here, again and preeminently, since reflection is the instrumentality of securing freer and more enduring goods, reflection is a unique intrinsic good. Its instrumental efficacy determines it to be a candidate for a distinctive position as an immediate good, since beyond other goods it has power of replenishment and fructification. In it, apparent good and real good enormously coincide" (Dewey 1925/1981, 303–304). The crucial point here is that reflection can be the "unique intrinsic good" because it represents the very process of critical inquiry that we need in order to adjudicate among competing values (Kahneman 2011). Critical reflection is, in a sense, the highest epistemic virtue—the only supreme virtue possible, once we abandon any notion of self-evident absolute truths and values (Johnson 2014a, chapters 5 and 8). It is our best means for sorting through competing values in order to discern their causal antecedents and their implications for future experience.

This conception of philosophy as critical evaluation of systems of values raises the issue of whether philosophy aspires to possession of higher truth, of ultimate knowledge of values (or of anything else, for that matter).

The answer is *no*. Philosophy explores *meanings* rather than amassing facts or grasping ultimate truths. Rorty (1979) correctly saw Dewey as challenging the reduction of philosophy to theory of knowledge. Rorty attributes this epistemologizing of philosophy to the fateful influence of Kant's idea of a "critical philosophy" that investigates the origins, nature, and limits of human knowledge (Kant 1781). When philosophy tries to become a competitor with science for accumulating truths about nature, mind, or thought, it will eventually get replaced by those sciences that have developed their powerful empirical methods and technologies of investigation.

Dewey argues that, *instead of pursuit of ultimate truth, philosophy's distinctive mission is the exploration of the meaning of different experiences, as the basis for living wisely*. Dewey's account of philosophy's empirical denotative method puts front and center the understanding of meanings:

> The realm of meanings is wider than that of true-and-false meanings; it is more urgent and more fertile. When the claim of meanings to truth enters in, then truth is indeed preeminent. But this fact is often confused with the idea that truth has a claim to enter everywhere; that it has monopolistic jurisdiction. Poetic meanings, moral meanings, a large part of the goods of life are matters of richness and freedom of meanings, rather than of truth; a large part of our life is carried on in a realm of meanings to which truth and falsity as such are irrelevant. And the claim of philosophy to reveal or displace science as a purveyor of truth seems to be mostly a compensatory gesture for failure to perform its proper task of liberating and clarifying meaning, including those scientifically authenticated. (Dewey 1925/1981, 307)

We cite this passage at length, because it sums up the entire direction of Dewey's argument, culminating in his conception of the nature and importance of philosophical understanding. Philosophy is pursuit of wisdom for living because it gives us an empirically grounded way to assess the meanings of systems of values that compete for our allegiance.

At least since the seventeenth century, philosophy has been in an increasingly anxious identity crisis. As the various sciences multiply, they gobble up ever larger slices of the pie of philosophical concerns. For example, in the nineteenth century, psychology began to separate itself from traditional natural philosophy and took with it the claim to experimental investigation of mind (James 1890; Wundt 1874/1893). In the late twentieth century, moral psychology takes over some of the more empirical parts of moral philosophy, so much so that it is often difficult to decide whether to characterize it as philosophy, psychology, or both (Greene 2014; Greene et al. 2001; Haidt 2012; Sinnott-Armstrong 2008; Solymosi & Shook 2014). At the same time, neuroscience is chewing up philosophy's prior exclusive claim to explaining perception, feeling, conceptualization, and reasoning (Churchland 1986; Damasio 2010; Edelman & Tononi 2000; Prinz 2004). As the sciences refine their methods and develop new experimental technologies, they tend to do a more empirically responsible job of explaining natural events than is possible via armchair speculation. It might even be said, without too much exaggeration, that logical empiricism, and analytic philosophy in general, originally emerged as attempts to find a legitimate role for philosophy that could not be taken over by the sciences. For example, philosophy was regarded by some philosophers not as science, but as the logic of science (Carnap 1928/1967; Hempel 1965; Popper 1959).

Dewey's version of this response to philosophy's identity crisis is to observe that knowledge acquisition is not the sole function of human cognition, but rather just one of many functions. There is plenty left for philosophy to do. There are realms of meaning that extend far beyond matters of knowledge and truth. We are not just *knowers*! We are makers, doers, players, communicators, lovers, and fighters. As Alexander (2013) reminds us, we are searchers for, and makers of, meanings

Metaphysics as the Ground-Map of Experiences

This recognition of the importance of meaning brings us back to the task of metaphysics, its relation to philosophy, and its use of empirical denotative method. One traditional recurring perspective takes metaphysics to be the ultimate exploration of the nature of being, not of individual beings, but rather of Being itself, or what Paul Tillich (1951–1963) called "the Being beyond all beings." Rejecting this conception of metaphysics as the exploration of ultimate reality, Dewey conceives the mission of metaphysics to be the description of the basic recurring characteristics of experience: "If philosophy be criticism, what is to be said of the relation of philosophy to metaphysics? For metaphysics, as a statement of the generic traits manifested by existences of all kinds without regard to the differentiation into physical and mental, seems to have nothing to do with criticism and choice. . . . When it has revealed the traits and characters that are sure to turn up in every universe of discourse, its work is done. . . . But the very nature of the traits discovered in every theme of discourse, since they are ineluctable traits of natural existence, forbids such a conclusion" (Dewey 1925/1981, 308).

As Rorty (1982) observed in his criticism of Dewey's notion of an empirical metaphysics, Dewey's suggestion that the generic traits are those "manifested by existences of all kinds" that occur in all universes of discourse, looks an awful lot like a traditional conception of metaphysics as setting out the basic categories of being. In his introduction to the *Later Works* edition of *Experience and Nature*, Sidney Hook suggested that, for Dewey, instead of positing a list of supposedly universal traits, "what he is really interested in is cataloging and analyzing those features of the world that have an important bearing on the human condition, on human hopes and possibilities, that are often taken for granted without being clearly articulated" (Dewey 1925/1981, xiv). Hook's suggestion seems to capture the spirit of what Dewey is really interested in directing our attention to, while yet avoiding any scent of an objectivist metaphysics (Hook 1927).

Dewey's "empirical" metaphysics is a description of the meaning of general patterns and qualities that recur in experience. Metaphysics thus provides the description (the denotation) of the generic traits that have to be considered for a suitably rich, deep, and comprehensive account of the happenings of nature as they bear on our lives. It gives philosophy its focal subject matter, as a basis for criticism of value systems.

Dewey never supplies any definitive list of what he takes the relevant generic traits of existences to be (see Alexander 2013, 108), but some of the ones he mentions throughout *Experience and Nature* are:

- situation
- quality
- the precarious and the stable
- constant relations
- continuity
- interaction
- temporality
- contingency and need
- movement and arrest
- novelty
- end-as-closure versus end as end-in-view

- means–consequences
- valuing and valuation

We can now appreciate the role of such traits in philosophy. If these truly are recurring general characteristics of experiences, then any philosophy that claims to explore and explain experience must take these (and possibly other) traits into consideration. Failure to do so results in the Philosopher's Fallacy of selecting some dimension (e.g., the stable aspects of a situation) and then pretending that it covers everything that is important in experience.

Metaphysics is thus the identification of these generic traits of experiences that structure the situations we find ourselves in, and it explores the meaning of these traits, with respect to both their causal antecedents and their consequences for future experience. This accounting of traits common to different kinds of experiences "is source both of values and of their precariousness; both of immediate possession which is casual and of reflection which is a precondition of secure attainment and appropriation. Any theory that detects and defines these traits is therefore but a ground-map of the province of criticism, establishing base lines to be employed in more intricate triangulations" (Dewey 1925/1981, 308–309; Fesmire 2015, chapter 2).

This brings us full circle to the nature and importance of an "empirical metaphysics" of the sort described earlier, in the chapter on experience and method. We saw that Rorty (1979, 1982) was right to reject the conception of metaphysics as the ultimate inquiry into the nature of being, with all of the foundationalist epistemology that goes along with that project. Nevertheless, thinking has to start *somewhere*, with some circumscribing of the relevant phenomena to be described, explored, and explained. Whatever conception of philosophy you propose, you have to decide what it is about—what it is attempting to illuminate. Therefore, you cannot avoid engaging in some form of empirical metaphysics. If the descriptions (the ground-map) are too selective or too reductive, then your philosophy is bound, to that extent, to be inadequate in precisely those respects.

One is always welcome to propose additional generic traits not previously identified, or to propose a reconfiguration of one's current ground-map, but then empirical responsibility requires you to explore the implications of so reconfiguring the map For example, let's say that you think of knowledge as concerned only with the stable and fixed aspects of nature. Then you will select only stability (and fixity) as the chief characteristic of all existences—that is, of what's *real*. But the result of this is that you then find yourself having to relegate to the status of nonbeing the innumerable occurrences of precariousness (and change) that you encounter every waking hour. You might wonder about the adequacy of any metaphysics (like yours) that has to treat change as either an inferior type of being, or a privation of being. History is not lacking in metaphysical systems that deny to change the status of genuine being, but, as we have seen, such theories that focus exclusively on the more stable dimensions end up distorting, mangling, and denying our experience of the precarious and changing dimensions which play such a huge role in our values, our thinking, and our actions.

Dewey's metaphysical posture, then, is that every philosophical theory must have *some* metaphysical ground-map. The inclination of many who have been called "linguistic" or "analytic" pragmatists or neopragmatists (Brandom 2011; Koopman 2009; 2013; Price 2013; Rorty 1982) is either to deny the need for any metaphysical commitment, or else to claim that an account can be given of the conditions for the successful performance of certain kinds of communally constituted acts, such as justifying claims about knowledge, truth, or values, but in a manner compatible with multiple different ontologies. This amounts to the claim that you can have a somewhat metaphysically neutral philosophy.

Considering this second thesis first, the claim that your favored analysis of community standards of thought, action, and valuing can be compatible with various metaphysical schemes, at least admits the necessity of affirming *some* metaphysical view. The first thesis—that philosophy can do its work absent *any* metaphysical commitments—is far more radical. We find this view incredible, at best, and irresponsible, at worst. When, in "Justice as Fairness: Political, not Metaphysical" (1985), John Rawls proposed a metaphysics-free moral (and political) philosophy, which he called "Kantian Constructivism" (Rawls 1980), he was immediately set upon by critics like Michael Sandel (1982), who pointed out metaphysical assumptions hidden in what Rawls called "the Original Position," from which a set of moral (or political justice) principles could supposedly be projected. Sandel and others showed how Rawls had presupposed certain characteristics of human beings, their motivations, the structure of their reasoning, and the environments they inhabit. So, there was a metaphysics there, even though Rawls denied it.

It may seem mundane, or even trivial, to say that human beings have bodies with certain needs, that human thought has certain body- and brain-based structures, that we can only live and act in certain types of environments, but these *are*, or imply, metaphysical claims. The postulation of a metaphysically neutral system of thought is just a bluff waiting to be called. If we can identify even one metaphysical commitment involved in your theory of, say, reason-giving or justification of truth claims, then the entire metaphysics-free philosophy edifice crumbles. Say, for instance, that, like Robert Brandom, you are concerned with the conditions for successfully performing certain kinds of speech acts (e.g., stating truths, making suggestions, giving orders, expressing feelings, instituting values). You have to assume that people exist, that they have the capacities necessary for certain communicative acts, that there are objects in the world to interact with, to set values on, and to communicate about. You might respond, "Well, of course, I'm assuming that there are such things as people, and they can do certain actions, and that they live in communities and within cultural systems, but I'm only interested in the conditions of performativity for their actions." OK, but performativity depends on a metaphysics, an account of the generic traits of existences. Performing a particular kind of speech act (e.g., stating, questioning, berating, complimenting, christening, condemning) depends on a number of assumptions about people and their surrounding world.

The point here is that everybody has his or her metaphysical perspective. The issue isn't whether you have a metaphysics (you do!). The issue is whether you have merely unreflectively accepted some culturally inherited metaphysical system (some particular ground-map of existence), or whether you have an appropriately comprehensive, rich, deep, and adequately critical metaphysical stance (Hildebrand 2019; Johnson 2014b). There is no need to be afraid of metaphysics, properly conceived. We have inherited a misleading dichotomy that posits only two radically opposed positions: either we must accept an absolutist (objectivist) metaphysics with its foundationalist epistemology, or else we must deny the need for any metaphysical stance whatever. Dewey, of course, rejects traditional objectivist metaphysical systems that posit Being (capital "B"). However, that sort of fixed and finished absolutist ontology is not the only option for metaphysics. Rejecting this radical forced choice between metaphysical objectivism and having no metaphysics at all, Dewey proposed his empirical metaphysics of the generic traits found in most experiences and therefore manifest in various universes of discourse. He acknowledges that, being an empirical undertaking, this metaphysics is subject to the same methods of testing and assessment that are appropriate for all empirical modes of inquiry: "The generic insight into existence which alone can define metaphysics in any empirically intelligible sense is itself an added fact of interaction, and is therefore subject to the same requirement of intelligence as any other natural occurrence: namely, inquiry into the bearings, leadings and consequences of what it discovers" (Dewey 1925/1981, 310). Such an empirical metaphysics is thus not only possible, but quite necessary once we have given up any claims to absolute knowledge. As long as we avoid a classical metaphysics of Being (of a fixed, completed universe), and as long as we recognize that any metaphysical ground-map might need to be revised in the face of changing conditions, there is nothing to fear here. Moreover, Dewey's generic traits of existences are so general and comprehensive that it is hard to imagine changes in experience so dramatic that such traits might disappear. Granted, what we can currently imagine is not the final arbiter of what is possible, so we should never terminate our metaphysical inquiries (Neville 1992).

Philosophy and Valuation

Philosophy is general criticism (valuation) of widespread systems of criticism. This raises the question of the origin and nature of values. Significant parts of many of Dewey's writings are dedicated to criticizing traditional theories that posit transcendent values, separated off from natural events, so that somehow we need to get those values back into relation with things in nature. This amounts to the view that norms (values, standards) are not *in* and *of* nature, yet they are supposed to be applicable to natural goings-on. Dewey observes that such a view has a *deus ex machina* character: "'Standards' of value suddenly make their appearance to serve as criteria of taste and conscience. The distinction between likings and that which is worth liking, between the desired and the desirable, between the is and the ought, descends out of the blue" (Dewey 1925/1981, 317). This separation of values from natural events does not go away merely by claiming that the values and standards are issuances of practical reason (Kant 1787/1993). This just puts reason in a transcendent realm of norms, a strategy that only redefines the problem as one concerning how a nonnatural (and normative) reason can be binding on a natural organism, here an embodied human being.

It is at this point that the radical character of Dewey's view of values becomes evident. From a naturalistic perspective, values, reason, meaning, thought, or knowledge cannot be separated from natural processes and events. "Such a theory will realize that the meanings which are termed ideal as truly as those which are termed sensuous are generated by existences; that as far as they continue in being they are sustained by events; that they are indications of the possibilities of existences, and are, therefore to be used as well as enjoyed; used to inspire action to procure and buttress their causal conditions" (Dewey 1925/1981, 311). In sum: *if you never start by taking values out of nature, you will never have the problem of how to get them back into nature, or to show how they apply to natural events.* Instead, your problem will be which, among all the competing values vying for realization, ought to be supported and strengthened in a particular context.

Two points stand out here for special attention. First, values are *in* nature, and they arise from our biological, interpersonal, and cultural makeup and situation. We don't have to worry so much (only maybe just a little) about where our values come from. There are values aplenty, set by our biological development of basic needs and our bodily and social capabilities relative to the affordances of our surroundings. They arise also from our need for nurturance, care, and love from others. They come from our cultural needs, practices, and institutions. We have many fine accounts, scientific and otherwise, of those values, their origins, and their implications (see, for example, Hinde 2002). What is more pressing for us is how to adjudicate among conflicting values, which is where philosophy comes in. This is no easy task.

The second point of emphasis concerns the continuity among levels of values. Dewey doesn't think that there are distinct kinds of experiences (e.g., aesthetic, moral, religious, political, economic) that require distinct kinds of judgments, but he does identify continuities among what we commonsensically think of as aesthetic, moral, and philosophic value systems. We have presented our case in this book (especially chapter 8) for the pervasiveness and importance of our aesthetic sensibilities at all levels of our functional organization. Dewey thus regards what we call aesthetic judgments as having standards and involving sensibilities that operate also in what we call moral judgments, and that are carried up even into the criticism of value systems that is philosophy.

One stunning implication is that *beliefs are values*. As Peirce (1878) realized, beliefs are not abstract mental objects, or even processes, but rather tendencies toward actions. To believe something is to be disposed toward actions implicated by whatever the belief is. If you believe that water freezes at 32°F, then, if you want ice, you will turn down the freezer setting below 32°F. If you want the soup thawed before dinner, you will take it out of the freezer to a warmer ambient condition and for an appropriate amount of time. The irritation of doubt gives way to reconstructive action via formation of belief dispositions. Beliefs are values, in the sense of being found, through experience and inquiry, good to believe, relative to certain circumstances and certain needs and desires. The process of moving from valuing to valuation thus applies to assessment of beliefs just as much as it applies to value judgments about art and morals: "The distinction between an immediate belief-value, which is but a challenge to inquiry, and an eventual object of belief that concludes critical inquiry, and has the value of fulfilling the causal relationships discovered, is made in the course of intelligent experiment. The result is a distinction between the apparent and the real good" (Dewey 1925/1981, 319). So, there is a continuous thread through these events: from biological regulation to aesthetic appreciation to moral deliberation and all the way up to evaluation of beliefs. It is one continuous process of valuation, but realized at different levels of functional organization.

Here, obviously, the "real" good is not some absolute or final ultimate good, but only "the better," as that has emerged through inquiry, and subject to future reassessment. The good is a relative matter, a matter of better and worse, compared with other candidate values in a particular context.

> To say that an object is a good may seem to be an absolute and intrinsic declaration particularly when the assertion is made in direct act rather than in thought. But when we recognize that in effect the assertion is that one thing is better than another thing, the issues shift to something comparative, relational, causal, intellectual and objective. *Immediately* nothing is better or worse than anything else; it is just what it is. Comparison is comparison of things, things in their efficacies, their promotions and hindrances. The better is that which will do more in the way of security, liberation and fecundity for other likings and values. (Dewey 1925/1981, 12)

This relative, comparative notion of value is perfectly compatible with what we know about animal learning. Animals have evolved capabilities for perception and action. A key value is always achieving some desired state, while adequately preserving resources, avoiding depletion of energy that might be needed for future tasks. Consequently, animals are not designed for perfection, but rather for doing what is *good enough*—good enough to secure food, locate water, find shelter, mate, and stay alive. The primary goal is viability, not perfectibility. The lion or cheetah will pursue the wildebeest just so long, until the energy expenditure is not worth the probability of actually getting that meal. Sometimes, it's better to save energy resources for the next waterhole, hoping to improve chances for an easier kill.

This "good enough" strategy has been handed down to humans, and it operates not just at the level of physical sustenance and survival, but all the way up to abstract conceptualization and reasoning as part of problem solving. On a timed logic examination, you may struggle mightily on a certain proof, without solving it, but, at some point, if you are prudent, you will abandon that proof and move on to the next, hoping for a more efficient and successful employment of mental resources. The punchline for this account of the "good enough" strategy was articulated in the film *Babe*, in the wise words from the farmer to his beloved pig: "That'll do, pig/ that'll do." Enough is enough when it's good enough.

Dewey elegantly ties up all of these characteristics of values in the following passage: "As inquiries which aim at knowledge start from pre-existent beliefs, so esthetic and moral criticism start from antecedent natural goods of contemplative enjoyment and social intercourse. Its purpose is to make it possible to like and choose knowingly and with meaning, instead of blindly. All criticism worthy of the title is but another name for that revealing discovery of conditions and consequences which enables liking, bias, interest to express themselves in responsible and informed ways instead of ignorantly and fatalistically" (Dewey 1925/1981, 321). Philosophy is this kind of activity carried up to the level of critical evaluation of systems of valuation.

When it comes to constructing an adequate account of valuing and valuation, the primary obstacle is the unreflective assumption of dichotomies like values versus natural existences. When you begin that way, you are already denying interaction as the progenitor of values. The temptation, so often succumbed to in theories of aesthetics, morality, and knowledge, is to divorce values from natural events, and thereby to deny interaction. The only satisfactory alternative is to recognize the natural intertwining of values and physical events. Interaction is the key to solving the problem of how values operate in our lives: "Yet if man is within nature, not a little god outside, and is within as a mode of energy inseparably connected with other modes, interaction is the one unescapable trait of every human concern; thinking, even philosophic thinking, is not exempt. This interaction is subject to partiality because the human factor has bent and bias. But partiality is not obnoxious just because it is partial. . . . What is obnoxious in partiality is due to the illusion that there are states and acts which are not also interactions" (Dewey 1925/1981, 324). Interaction—from the beginning of life—means that our perceptions, actions, values, thoughts, and intellectual achievements are natural goings-on, and not parachuted into the world from a pure realm of the intellect, or a pure realm of norms. We live in, among, and through those intertwined, sometimes conflicting and sometimes harmonizing, values. Our challenge is to intelligently reorder and reconstruct some of those values on the fly, struggling to resolve conflicting values right in the middle of things, as we are inescapably caught up in the ongoing flow of experiences.

An account of Dewey's moral philosophy obviously deserves its own chapter, or a whole book (see, for example, Fesmire 2003; Johnson 2014a; Kitcher 2011), but we will settle here for one general statement about the character of such a moral theory. First, all moral philosophies—whether deontological, consequentialist, or virtue theories—that select but one dimension of human moral cognition (e.g., divine commandments, reason, the good, virtue, moral law, or a moral calculus of goods) are bound, from the start, to be inadequate by virtue of narrowly selecting that one factor of human morality and attempting to erect it into a grand narrative of all morality (Dewey 1922/1988; 1930b/1988). For example, any deontological theory that claims to discover universally binding absolute moral principles derived either from divine mind or universal reason will not be adequate in the face of newly emerging conditions different from those in which the ethical principle or moral law was first formulated. It will tend to overemphasize inner resolve (as goodwill) and adherence to principle, while undervaluing consequences of actions. But utilitarian ethics has its own problems, when it claims that we can construct a moral calculus to assess the goodness or badness of proposed actions, for this strategy ignores the complexity and changing qualities of a morally problematic situation. It will try to substitute calculation for sensitive discernment of what best resolves value conflicts. Another one-factor view, such as virtue theory, will overemphasize dispositions to act, at the expense of not fully appreciating the goods to be realized through action. The exclusive selection of particular phenomena and key values by these three types of theories is an example of the Philosophic Fallacy of selecting one dimension and claiming that it is the one essential factor capable of resolving all moral problems. As we would expect, Dewey's assessment is that *all* of these factors are relevant and must be considered, but there are no ultimate rules or algorithmic procedures for weaving these dimensions together to generate a moral judgment (Dewey 1930b/1988).

As we saw in chapter 7, Dewey proposes the notion of moral deliberation as a dramatic rehearsal in imagining the probable consequences of acting on a certain value or set of values within a particular situation and then assessing which values, if promoted, would better resolve the problematic situation. Dewey accuses most traditional moral theory of assuming the separation of physical events and values, and then having to bring those values "down from above" to bear on a concrete situation. This strategy assumes that values are given and then need to be applied to experience. One version (deontology) would say, "Here are the moral imperatives, now go and apply them to life." Another (the consequentialist) might say, "Here is the mechanism for assessing the amount of well-being that will result from specific actions, now go and use that tool to calculate the correct response." And the virtue ethicist might say, "We [your family and culture] have cultivated the proper virtues in you, now go and express them in morally right actions."

Dewey's posture would be more like, "Each of these theories provides some insight into important factors in moral valuation. But what is needed is to employ your capacities for criticism to inquire into the meaning (i.e., sources and consequences) of all the varied systems of values (theological, deontological, consequentialist, virtue-based, ecological, feminist, etc.) that you can find, and then use this understanding, via dramatic imaginative rehearsal, to create a more harmonious, liberating, and expansive way of being in the world, here and now." Instead of claiming to be the discoverer of the ultimate moral values, ethics is that part of philosophy-as-criticism that focuses on issues of better and worse by way of human conduct. From this perspective, a suitably reflective moral agent would realize that, while all long-lasting moral systems provide important insight, you should not try to reduce moral judgment to a single factor, such as rule following, quasi-mathematical calculation, or development of dispositions without attention to surrounding conditions and other significant factors (such as consequences of certain acts). In sum: "The business of moral theory is not at all with consummations and goods as such, but with discovery of the conditions and consequences of their appearance, a work which is factual and analytic, not dialectic, hortatory, nor prescriptive. . . . The office of moral philosophy is criticism; and . . . the performance of this office by discovery of existential conditions and consequences involves a qualitative transformation, a re-making in subsequent action which experimentally tests the conclusions of theory" (Dewey 1925/1981, 323).

When it is realized that values (moral, aesthetic, and otherwise) live in nature, not outside it, we come to see values as pivots for the (hopefully) intelligent transformation of experience, for the better. Values are not outside us, in the sense of dropping down from a transcendent realm of norms. We are in and with them in inescapable interaction with our surroundings. We move through them, under their forceful influence, relying on our modest capacity to rework them toward desired, more satisfactory, more harmonious ends-in-view. Our best tool in all of this is our native and acquired intelligence: "Because intelligence is critical method applied to goods of belief, appreciation and conduct, so as to construct freer and more secure goods, turning assent and assertion into free communication of shareable meanings; turning feeling into ordered and liberal sense, turning reaction into response, it is the reasonable object of our deepest faith and loyalty, the stay and support of all reasonable hopes" (Dewey 1925/1981, 325). Philosophy is this intelligence operating at the highest levels of generality in assessing the meaning and limitations of major systems of values.

10 Living with Naturalism

In his *Lectures on Logic* (1800/2004), Immanuel Kant identified four major questions that he thought a philosophy worth living by ought to answer: (1) What is man? (2) What may I know? (3) What ought I to do? and (4) What may I hope? Even today this remains a good initial short list of some major issues we want a philosophy to investigate, though it is only a starting point. In this closing chapter, we propose to use these questions, suitably rephrased, as a way to address the nagging issue of why we should care about a naturalistic perspective of the sort we have been developing here. In short, why bother? What conceivable difference could and should it make in our lives?

It should be clear by now that we believe there are strong reasons for adopting a naturalistic philosophy, and that such a view supplies a meaningful philosophy for living. It enlightens us about who we are, what our capabilities are, and what we can aspire to. It gives guidance for moral deliberation. It bases this account on a scientifically sound explanation of what makes us what and who we are, and it uses this self-understanding to provide wisdom for living.

Some people are put off by the very idea of a naturalistic philosophy. They see it as emphasizing our biological, animal nature, while downplaying our intellectual, spiritual, moral, and aesthetic capacities. Like the 1950s vocalist Peggy Lee, they ask "Is that all there is?"—are we merely animals clawing our way through life, subject to evolutionary forces, until our biological clock stops, and we die?

Well, yes, we *are* suggesting that there is currently no respectable empirical evidence for either a disembodied soul or a transcendent ego, while there are mountains of scientific evidence indicating that humans are complex social animals, and that all our capabilities, from processes of feeling, perceiving, and moving, up to our highest intellectual, moral, and aesthetic achievements, are rooted in our animal engagement with our surroundings. We suggest that the picture of humanity provided by a naturalistic orientation of the sort developed here gives us a very attractive, realistic, and helpful philosophy worth living by. Such a view recognizes our natural capacities for perceiving, moving, feeling, thinking, and valuing which make it possible for us to inquire into our natural and social worlds, and to manage our values wisely and well. It is a humble philosophy, but one that provides practical guidance for human flourishing.

Under a slightly revised, more politically correct, version of Kant's four questions, we propose to summarize the key themes of this book and to explore their important implications for how we ought to view ourselves and how we ought to live. The subtitle of the 1964 film *Dr. Strangelove* is *How I Stopped Worrying and Learned to Love the Bomb.* On a far more optimistic note, the subtitle to our book might be *How I Stopped Worrying and Learned to Love Naturalism.* We wouldn't expect everyone to *love* naturalism, but we do hope to have made a good case for regarding it as a philosophical perspective worth living by.

What Is (Hu)man?

Humans are evolved and evolving complex social animals. It seems obvious to say that humans are animals, but it bears repeating, because over and over again, across the past three thousand years, the idea that we are *rational* animals has seemed to many people to indicate that we are somehow not just animals, but something more. According to this view, it is our rationality that defines us, not our animality, and far too often this rationality is taken to be independent of our animal nature. We have been arguing that our animality and our embodiment need to be recognized as the basis for all our capacities, including those we share with other animals and those which appear to be somewhat unique to humans (e.g., abstract thought, complex language, impulse control, and deliberative capacity). Our embodiment colors and shapes everything we experience, think, value, and do.

Before we get too impressed with our species, however, it behooves us to remember that, in some ways, we are not as good as certain other species of animal at specific tasks. We can't run as fast as cheetahs, stay submerged in water for long periods the way fish can, or fly unassisted like the birds, to name but a few constraints. Honest humility about our capabilities is a nontrivial virtue, and yet, despite these and many other human limitations, we are nevertheless special in certain important respects. Especially by virtue of our evolved capacities for experiencing, making, and communicating meanings, we have cognitive resources not found in the same degree, and with the same complexity of organization, in other animals. As far as we know, we are the only creatures capable of abstract conceptualization and reasoning, tied to advanced use of language and other forms of symbolic interaction, and the only species so far capable of reflectively controlling our actions, though only in a very modest way (Marler 1961; Marler & Hamilton 1966). So, we are animals, albeit very complex social animals, sharing some of our sensory, motor, and rudimentary intellectual and moral capacities with other animals, but also capable of certain functions not found elsewhere in the animal kingdom. That said, all claims about human distinctiveness need to be tempered by attention to evolving scientific research on what are very sophisticated capacities of various animal species.

The second crucial tenet (the first being our animality) is that everything about us stems from *interaction*—the ongoing interaction of our organic biological and social selves with our material, interpersonal, and cultural surroundings (Schulkin 2000). We exist only in and through our continual engagement with our environment. Evolution typically supplies us at birth with a more or less species-wide brain and body functional architecture, although there may be considerable intraspecies variations in how our brains are functionally structured. Connections among these functional regions are then created and tuned up through our individual developmental experience over the course of our lives (McEwen 1998, 2007). The patterns of neural connectivity and activation depend on the current makeup of the organism that is responding to the affordances of its surroundings (Gibson 1979). No interaction, no organism. No interaction, no experience. That is why Dewey says that "interaction is the one inescapable trait of every human concern" (Dewey 1925/1981, 324).

What makes recognition of organism–environment interaction so profoundly significant is its implication that everything in experience has to emerge within and from natural interactions. There is no place for causes, forces, ideas, motives, or values to drop down into experience from some allegedly supernatural or transcendent realm. One of the central guiding themes we have been developing throughout this book is that we humans are always *in*, *of*, and *through* natural goings-on. We are always in the thick of things, in continual interaction shaped by our various capacities, needs, values, projects, and environments. We are "*in it*" (i.e., in nature) through and through, just as long as we are alive. The real question is whether we are in nature reflectively, knowingly, intelligently, and wisely, or whether we are floundering in it blindly, ignorantly, and haphazardly.

The third implication is that interaction entails that we are selves-in-process. We are always "on the way," not toward some state of perfection, or realization of a pregiven essence, but on the way to wherever and whatever, evolutionarily and developmentally, we happen to be on the way to. We saw in chapter 7 that the self is an interpenetration of habits that constitute character and will. The self as currently formed is the result of these prior habits carried forward into the present moment, but also always opening out toward future possibilities. In this way, the self is never completely fixed and finished, but rather malleable, permeable to new events, and subject to transformation. The present self is a tentative relative stability, with good enough stability (via homeostasis and allostasis) within an ongoing flow of natural processes (Schulkin 2003; Sterling 2020).

This insight is a big deal. One of the great philosophical fallacies is hypostatization or reification of what is a functional process into an alleged object or cause of some sort. A primary instance of this fallacy is the hypostatization of so-called mental faculties. Given our various sensory, motor, and cognitive activities, we have a tendency to regard those functional organizations as fixed, discrete faculties, such as sensation, feeling, imagination, understanding, reason, and will. We then end up postulating a set of faculties that supposedly generate specific types of judgments. For example, based on our recognition of logical or epistemic relations among things and events, we mistakenly jump to the postulation of a preexisting essential structure of reason that supposedly makes it possible for us to know and reason about the world. How easily we seem to do this as we look to give ourselves foundations for various processes and capabilities. According to this mistaken view, being rational then consists of conforming our thinking to some pregiven rational structure (Fodor 1976). Another prevalent example is turning our knowing engagement with our world (which is a functional *process*) into a fixed *product*, knowledge. This view often carries with it the assumption of some inner homunculus or a transcendent ego that does the knowing (Dewey 1929/1984; Johnson & Tucker 2021).

A related hypostatization moves from the fact that we are creatures who conceptualize, to the claim that there must be quasi-entities known as concepts and propositions, that exist within something called mind (which, note well, is just another basic objectification of a functional process). These abstract conceptual and propositional entities then either do or do not correspond to mind-independent realities, and we are thus off and running with the skeptical problem of how what is inner (concepts, propositions, thoughts) can possibly relate to what is outer (things and events in the world). It is then easy to become hopelessly lost in reflections on what it is about a concept or a thought that guarantees its correspondence to a mind-independent world. We can then lose sight of the fact that knowing is an activity and a functional process of experiential transformation, not an inner mirroring of an external world.

It is worth reiterating, one last time, just how insidious and profoundly problematic dualistic metaphysics and epistemology are. Everywhere you look, you can find mistaken dualisms that have worked their way into every nook and cranny of our lives: mind/body, mental/physical, cognitive/emotive, conception/perception, reason/emotion, thought/feeling, knowledge/imagination, fact/value, public/private, and on and on. Once such dualisms are set in stone, it comes to seem obvious that we need to explain actions as the result of ideas, ends, motives, and values that are outside of natural events, but are then brought (externally) to bear on those events. A commonplace example of this is the alleged *is–ought* distinction, according to which nature is a set of causally interactive, value-neutral objects and events, and then values are brought in from some other source, to be applied to natural goings-on.

These dichotomies have become so definitive of how we think and act that we are often not even conscious of their working. We take them as self-evident givens—as built into the nature of things, and thus as calling for explanation in their own terms. They have become taken-for-granted distinctions that shape our experience, thought, values, institutions, and communal practices. Even when we are entertaining a naturalistic explanation, we tend to slip back into dualistic terms. For example, the naturalistic thinker wants to emphasize mind and body as aspects of an ongoing interactive process, showing how "mind" and "body" are one. However, even in asking how mind and body can be part of an interaction, we have already spoken of two separate things, mind and body, that somehow need to be brought back together again. In short, we have activated a pernicious mind–body frame that colors all our thinking.

Because we are so thoroughly influenced, unreflectively, by such dualistic distinctions, when somebody comes along with a nondualistic, naturalistic perspective of the sort we have developed here, people often have an intuitive knee-jerk reaction against any form of explanation that doesn't traffic in their preferred dualism. This leads to the common objection that naturalistic views try to reduce one side of a taken-for-granted mind–body dualism to the other side, typically reduction of mind to body.

The fundamental error here—and it is a huge error—is to interpret any dualism as indicating two distinct and independent structures or processes. Ontologically, this error amounts to the claim that the two parts of the dualism (here, mind and body) are two completely different kinds of things or events. It is then a short step from positing two different ontological kinds to the idea that each kind requires its own distinctive explanatory framework. The standard story is that physical objects and events require *causal* explanations, whereas mental phenomena, which concern thought, meaning, reasoning, and willing, require a different, *interpretive* method that allows us to appreciate the meaning of mental events and to recognize some role for freedom of the will. This methodological dualism then usually ends up denying the "natural" sciences (and sometimes even the social sciences) any significant role in the explanation of human meaning and values.

We have suggested that Dewey was right when he observed that the poles of any dualism indicate dimensions or characteristics of a continuum of functional processes, and *not* distinct ontological kinds of things. The dualistic error is thus the turning of what is a functional organization of organism–environment interactions into metaphysical entities or causes. Once you do that, you become locked into a dualism from which you can never extricate yourself, insofar as the very terms by which you have defined the problem to be solved assume a radical metaphysical and epistemological dichotomy.

The nondualistic alternative to this posture is to recognize the primacy of organism–environment *interaction*, and thus the primacy of ongoing activity and development, instead of fixed structures. One very useful strategy for understanding Dewey's philosophy is to take any noun, especially any noun for an abstract concept, and turn it into a verbal form to get Dewey's process (or event) metaphysics. This works, simply because by replacing a noun with a verb or adverb, it reverses the hypostatization by which functional processes are turned into either preexisting structures or finished products (i.e., quasi *things*). Following this noun-to-verb translation, instead of knowledge, it's know*ing*; instead of reason, it's reason*ing* and reason*able*; instead of conscience, it's *conscientiousness*; and instead of value, it's *valuing* and *valuation*. This move brings back into experience pervasive change, process, and activity. It allows us to replace ontological and epistemological dualisms with functional accounts of mind and thought. Recognition of interaction emphasizes activity and reinstates functions in their proper realm of activities and values.

And so, imagine a philosophy (and metaphysics) that took this advice to heart. It would see everything as processes and events—interactions occurring in and through time. It would keep you from turning functional activity into fixed states of being. It would not allow you to explain natural human events as the result of grand hypostatized "causes" like greed, fear, libido, anger, or even self-interest. Causation is not one unified thing, but many. There are no monolithic causal forces, but rather complex interweavings of multiple enabling conditions and contextual influences. Dewey (1922/1988) pointed out the ridiculousness of attributing war to some single "aggressive instinct" (as in, "Question: Why do we go to war? Answer: Because we have the warring (aggressive) instinct!"). He pointed out that there are as many types of aggression as there are objects and contexts of aggressive behavior. There is aggression toward military foes, resistance to bullies, fighting back against racism, competition among athletes in sporting events, and competition for political victory. There is sexual aggression, rough and tumble play among children, and aggressive activity appropriate for games of all sorts (Kagan 1984). There are these and many more manifestations of aggression, but there is no one chemical reaction that signals only aggression, and no isolated circuit in the brain dedicated only for aggression (Valenstein 2006). There is no single homogeneous aggressive instinct causing all these diverse aggressive behaviors, despite what some have thought (Lehrman 1953).

However, even though there is no single instinct underlying them, aggressive events are not merely social constructions. There are, indeed, tendencies activated within suitable contexts in which malicious aggression is readily expressed. Our aggressive transgressions are real and fill the litany of our history of conflict and violence across the globe and over the course of history.

Dewey frequently observed how giving up the objectification of grand causal forces and grand instincts would dramatically change our whole way of approaching causal explanation, making possible a more scientifically correct and nuanced orientation. We could then see that causation is complex, contextual, and always relative to our interests and values. It is not one thing, but many diverse, though related, enabling conditions operating together. What we call "the cause" is a selection of conditions that we find useful for explaining and intervening in certain events. For example, consider how our ignorance of science so often leads us to fail to recognize causal complexity, that is, the interaction and interdependence of multiple conditions, leading to multiple consequences. A particularly relevant example is the misunderstanding of global warming. Climate change deniers are sometimes heard to say things like "How can there be global warming, when last winter was the coldest on record here in the Midwest?" The answer is that global warming is not merely a one-factor causal phenomenon, but rather the result of multiple interacting conditions over an extended period of time, issuing in multiple diverse consequences.

Interaction is thus one of the great ideas of a naturalistic philosophy. It gives us a profoundly nondualistic picture of human nature. It requires us to rethink our inherited notions of explanation that rest on hypostatization. It reminds us that coming to grips with continual change is one of our great existential and philosophical challenges. We have to see ourselves as changing and always in process. We have to acknowledge that we are not the endpoint or goal of evolution. Evolution has no goal. We need to realize, as Nietzsche tried to teach us, that "man" (the human animal) is not the final end of evolution, but only just a stage on the way to something else, not something necessarily higher in capacities, but just different, in response to changed conditions. However, contra Nietzsche, there is no *ubermensch* to be realized as the "highest type." There is only us as we are now. Later, there will be something else—us as we are then. And this process of change will carry forward toward an indeterminate future of steps on the way to who knows what.

What Can I Know?

The main thesis of chapter 6 was that knowing (a verb for an activity) is the need–search–satisfaction process when it resolves (in some measure) a problematic situation. The need–search–satisfaction process emerges first in our biological functioning and is key to the maintenance of homeostasis and allostasis in our internal milieu in response to our changing physical, interpersonal, and cultural environments. Preserving continuity of functions, this three-phase process operates also at the highest levels of abstract conceptualization and reasoning (Johnson & Tucker 2021). There is thus a fundamental continuity between so-called lower and higher levels of functional organization. What we think of as higher cognition has the advantage of consciousness, meanings, mind, and language, so knowing at that level of organization is potentially more extensive, complex, rich in meaning, and flexible than problem solving within our biological processes. However, at all levels of functioning we find the need–search–satisfaction structure, though how that structure is enacted will vary according to the context in which it is operating. Knowing is thus an emergent function. Seeing it as such is the only way to avoid the dualism of mind and body, cognition and emotion, knowledge and belief (Crippen & Schulkin 2020).

Once again, we are brought back to the inescapable workings of interaction. Knowing has to find its place in the patterns and qualities of organism–environment interactions. We cannot stand outside or above our experience and look down on it from some transcendent perspective, as if to discern how a sentence or proposition maps onto a mind-independent state of affairs. To repeat the mantra, *we are always in the thick of things*. "The organs, instrumentalities and operations of knowing are inside nature not outside. Hence, they are changes of what previously existed: the object of knowledge is a constructed, existentially produced, object" (Dewey 1929/1984, 168).

Knowing is thus an activity that transforms experience, and it is a completely natural process of animal problem solving. It is not a process of metaphorical seeing, as in what Lakoff and Johnson (1980, 1999) have named the Knowing Is Seeing metaphor, which generates expressions like "I finally *saw* how to solve the equation," "Could you *shed more light* on what you just said?" "That was an *illuminating* argument," and "A more *obscure* argument would be hard to find." Dewey rejected the Spectator Theory of Knowledge, according to which some inner homunculus supposedly "sees" pregiven objects (ideas, concepts, propositions) and their abstract relations. In sharp contrast to the Spectator Theory, Dewey understood knowing as more a *doing*, an active reconstruction of experience through body-based inquiry. Knowing is known by what it does for you, that is, how it allows you to go forward and be at home in your surroundings, not for all time, but at least under relatively stable current conditions. If existential conditions change, then we need new adaptive inquiry.

What we can know is thus a product of our bodily makeup and capacities as they engage the affordances of things and events in the world. Up to a certain point, Kant (1781) got this partly right. He showed how what we can know, and how we know it, is the result of both what we humans bring to bear from our sensory, motor, and conceptual capacities, and also the patterns of energy we encounter in our environment. Unfortunately, he thought that relativizing knowledge to things as they appear to us was a concession to skepticism concerning any knowledge of things as they are "in-themselves" (i.e., noumena). Kant argued that there are some things (i.e., phenomena) we can know (as appearances), but we can never plumb the depths of things as they *really* are. This amounts to the deflationary skeptical claim that we think there is something more "out there" in the mind-independent world of things-in-themselves, but we'll never access it, and we are forever doomed to wallow in ignorance of the ultimate realities. We are left with a fundamental alienation of the organism from its world (as rooted in a noumenal realm).

Pragmatism frees us from such skeptical despair by abandoning the need for grasp of some ultimate, but forever hidden, reality. There is no point in, no evidence of, and no necessity for postulating unknown and unknowable things-in-themselves. Since we start with interaction, our knowing is rooted in the goings-on within our world. Interaction is between an organism and its environment, so the knowledge that arises within and from that interactive process is empirical, somewhat testable, and mostly reliable, though subject to error and thus requiring ongoing correction and improvement. Recall Dewey's remark that, since we are formed up in and through our ongoing relations with parts of our environment that sustain and challenge us, it is no miracle that parts of our environment are knowable in a way suited to our interactions, interests, and values: "The world is subject-matter for knowledge, because mind has developed *in* that world; a body-mind whose structures have developed according to the structure of the world in which it exists, will naturally find some of its structures to be concordant and congenial with nature, and some phases of nature with itself" (Dewey 1925/1981, 211).

What counts as knowledge is always dependent upon values deemed relevant, in a particular context of inquiry, and by a particular community of inquirers who provide critical perspective and enforce communal values of investigation and explanation. This keeps us from falling into a skeptical solipsistic abyss. We are subject to the critical scrutiny of other inquirers and can be held accountable for our failures of method, testing, and insight. However, knowledge is not just whatever a certain community says it is. That kind of relativistic view contributed to the burning of "witches," the genocide of indigenous groups in the Americas, the "final solution" during the Second World War, and the lynching of blacks in the United States. The pragmatist posture is rather that we should always seek to expand and diversify our perspective, as that is the only way to reveal the biases, prejudices, and narrowness of our present point of view, and to envision alternative understandings. This never finished attempt to extend the scope of relevant perspectives and the range of people whose experiences and interests get figured into our inquiries, involves an ongoing effort to increase what Peter Singer (1981) calls "the expanding circle" of lives, values, and views that count as relevant and important (Gilligan 1982). Consequently, instead of a self-satisfied and unreflective avoidance of critical input, we should, instead, submit our claims to knowledge to the critical scrutiny of other communities of inquirers.

Knowledge is therefore a relational, but not a completely relativistic, notion. There is no knowledge pure, absolute, and for all time. Nothing is immune to possible critique, but, as Quine (1951) pointed out, some views are so entrenched and so continually reinforced by recurring experiences that they may *seem* to be absolute and unchanging truths. They gain a relative stability and uncritical acceptance, and so they typically go unquestioned, even though they are not unquestionable. Kuhn (1962) showed us that nothing is intrinsically immune to scrutiny and challenge, though there are always beliefs taken for granted at any one time, in order to get an inquiry off the ground. We have to start somewhere, if we hope to get inquiry going. Those starting points are provisional, and they are subject to subsequent scrutiny and potential revision. Facts are therefore posits—hypotheses—for getting our inquiries underway in a meaningful and useful fashion. Facts are hypotheticals: If we assume X (the postulated fact), then we should expect Y, under appropriate conditions, to appear in experience. Therefore, we ought to investigate what is forthcoming in experience, and whether the postulated fact holds up under continual inquiry, or whether it needs to be revised or even rejected.

There is thus another big lesson here. We are not the kind of creatures who could have a god's-eye view of things. We only know things as they are meaningful to us, relative to our embodiment (i.e., our sensory, motor, and affective processes), dependent on our biological and sociocultural values, and acquired via methods, tools, and concepts (as technologies) that have historically arisen within communities of inquirers and practitioners. We don't and can't know things-in-themselves, but that's okay, because there is no reason to believe there are such unfathomable things. There is, and could be, no empirical evidence for so-called things in themselves, and we don't need any knowledge of that sort to get on with life. We only need an understanding of things and states of affairs as we engage them via various modes of bodily exploration, reflective inquiry, and other activities. We don't need ultimate reality, grasp of being, or intuition of absolute values. What we need are methods of inquiry good enough for our purposes. We need methods that have hypothesis projections, which are then subject to testing, criticism, and reconstruction of experience built into them. They will be fallible and subject to reconsideration in light of newly emerging conditions and values.

Moreover, an individual knower is not the final arbiter of what is good enough. What counts as warranted is a more objective consideration to be worked out in the broader communal context of inquirers employing multiple methods and subjecting hypotheses to critical scrutiny. Understanding and knowing are therefore relational activities, but they are not merely relativistic in the skeptical sense of that term.

Because all of our methods of inquiry and our conceptual systems are based on criticizable values, we have to give up any notion of absolute knowledge. Despite our pretensions to ultimate truths, we cannot be know-it-alls, but we can have confidence in the possibility of knowledge adequate to our current situation, and somewhat testable at our sensory and motor interface with our surroundings, especially when aided by scientific technologies (Johnson & Tucker 2021). The replacement for allegedly fixed and completed knowledge must be constant vigilance; that is, continued criticism of our systems of valuation (including belief values), with openness to both supportive and falsifying evidence (Popper 1959). Science is not one thing, but many methods and practices, yet we can say that what makes science a preferred mode of empirical inquiry is that it has built into its procedures various methods of criticism and revision, subject to evaluation by ever expanding communities of inquiry. There is no guarantee that some people won't cheat, or just not be very good inquirers (Pashler & Harris 2012), but empirical critical method is, overall, our best shot at an appropriate and workable system of investigation and criticism.

Our posture toward the sciences ought to be appreciation of the power of their impressive methodologies and technologies of investigation, but this embrace of scientific methods ought to be coupled with the need for ongoing scrutiny and criticism of the assumptions, limitations, and underlying values of various empirical methods. Sometimes we need to get our own house in order, reminding ourselves of our deep, unreflective biases. Confirmation bias—looking only for evidence that supports our preferred hypothesis—is a powerful tendency, and the only way to lessen its effects is the vigilance and critical perspective of communities of inquirers capable of seeing beyond, and taking issue with, our own narrow circle of assumptions and practices. Science is not, by any means, the be-all and end-all of knowledge, but the cluster of empirical methods practiced in the various sciences still remain our most robust ways of checking our own errors and faults, and growing knowledge in an empirically responsible way.

Because Dewey's account of knowledge focuses on *activity* as the basis for the transformation of experience through inquiry, Dewey is sometimes criticized for overemphasizing action at the expense of contemplation. His insistence on the central role of activity in knowing is a reaction against the Spectator Theory of Knowledge prominent in his day, and still popular today. The Spectator Theory conceives of knowledge as contemplation of fixed and completed objects, properties, relations, and events. Dewey, as we have seen, replaces the mind-as-mirror-of-nature metaphor with a focus on the importance of the search and foraging activity that structures inquiry, from the biological level all the way up to the level of abstract conceptualization and reasoning. Dewey needed to stress activity and making, in order to counteract the overemphasis on contemplation of fixed essences or kinds that characterizes so much theory of knowledge.

So, Dewey's is a philosophy of activity, but it is not exclusively about activity. If it were, it would be just another selecting of one pole of a duality, here the active over the passive, and would miss the crucial role of the consummatory in experience. In *Art as Experience* Dewey thus insists that "*an* experience" is a consummation of meaning that requires a delicate balance of doing and undergoing. Excessive doing (activity) keeps us always rushing forward, focused on controlling what happens, and it thus drastically limits our ability to savor the meaning of what is happening or being done—to smell the roses, so to speak. At the other extreme, excessive undergoing (passivity) lets experiences wash over us (as passive recipients of sensations), but we are not actively engaged with what is happening, so we miss important dimensions of the meaning of the event-in-process. Dewey concludes that the same play of activity and receptivity that defines our aesthetic sensibilities operates also in our everyday experience, on those occasions when it achieves meaning for us: "In short, art, in its form, unites the very same relation of doing and undergoing, outgoing and incoming energy, that makes an experience to be an experience" (Dewey 1934a/1987, 48).

The remedy to selective emphasis on just one dimension—doing or undergoing—is to let our appreciation of the meaning that is developing in a situation guide our search activity as we explore our present situation and its implications for further experience. The Buddhist call to "Be present to experience" would, in Dewey's framework, consist in savoring the emerging meaning arising via your engagement with your surroundings. It is an *active* being-in-the-world, a way of grasping the meaning of what is happening here and now. Knowing activity is thus a work(*ing*) of art that combines active search and molding of experience with attentive appreciation of what is unfolding before you.

What Ought I to Do?

We have already, toward the end of chapter 9, summarized a few key implications of a naturalistic orientation toward morality. Some of those are so important that it bears repeating just how consequential they are. For Dewey, it is a somewhat easy segue from "What can I know?" to "What ought I to do?" because all knowing is a mode of doing and making.

The first point to be underscored is that values are *in* and *of* nature. They are not extraterrestrial entities or forces brought into the world of natural events from some supposedly separate realm of norms. Instead, our values arise within the interactions we have with our biological, interpersonal, and cultural environments, based on our striving for viability and flourishing. Values arise through organism–environment interactions at all levels of our functional organization. The requirements for life-maintenance define basic biological values, and our primary intersubjectivity and need for nurturance generates interpersonal and social values. This multiplicity of values generates the conditions for the frequent conflict of values, including internal conflict (e.g., your desire for certain sweet and fatty foods, preferably in large quantities, versus your desire for a healthy, fit body) and also external conflict of our values with those of others (e.g., one person's economic desire for extraction of resources versus another's vision for a sustainable ecology).

Confronted with all of these potentially or actually conflicting values, we have to choose. We have to decide what we think is better and worse in the particular problematic situation we find ourselves in. Dewey (1922/1988) argued that whenever choices of better and worse arise, we are dealing with moral judgments, because issues of well-being are at stake. We saw that he rejects any notion of distinct types of experience, as well as the notion of correlative types of judgment as required by those types of experience. Consequently, all conduct is potentially a matter for moral assessment. Whenever the issue of better and worse arises—whether regarding works of art and aesthetic sensibilities, matters of explicitly moral concern, or decisions about belief and knowledge, we are engaged in moral deliberations. Therefore, whenever values conflict within or across any of his three plateaus of functional organization (i.e., physical, psychophysical, or mind), we are deciding what is better and worse, and we are thus engaged in moral evaluation (i.e., criticism).

In the face of value conflict, how do we choose which to prefer and nurture? Most people want a standard by which to rank order preferences to resolve the conflict. Traditional Western moral philosophy has offered three major candidates (with subvariations) for the ultimate decision procedure and standards of value. Deontological theories claim to identify ultimate principles or laws of duty (derivable from divine commandments, universal reason, or some communal source) concerning what, in a specific case of conflict of values, is morally forbidden, what is morally permitted (though not categorically required), and what is unconditionally required (Donagan 1977). The focus is on unconditional duties and the corresponding rights of moral agents to be treated with respect and fairness. Morally right action is seen as dependent on purity of will, the internal cultivation of a moral will capable of discerning our duty and having the willpower to do it.

The second basic orientation, consequentialism, counters that results, not just internal states of willing, matter, too, so it makes all the difference which goods you realize through your actions, and not merely how inwardly pure and morally strong you might be. Good intentions (good will) are important, but no more important than good actions that change experience for the better. Morality is about making the world a better place through actions that realize certain states of affairs regarded, via some standard, as good. Consequentialist theories can be either egoistic, focusing on maximizing the individual's perceived well-being, or utilitarian, focusing on maximizing the well-being of larger communities.

The third approach, virtue theories, recognizes both the importance of cultivating moral sensibilities and habits (as virtues) and appreciating their efficacy in realizing certain desired states of affairs. They emphasize the key role of the community in determining which dispositions and character formations are to be approved and cultivated for the well-being of the individual and society.

As discussed in the previous chapter, Dewey saw that we cannot reduce any of these three factors to the others. They are, to cite the title of his 1930 essay, "Three Independent Factors in Morals." He gives detailed explanations for why reduction to a single factor (e.g., duty, good, virtue) is neither possible nor morally useful (1922/1988; Dewey & Tufts 1932/1989). He also points out that many of our moral conflicts are the result of having to choose which of the three factors to prefer in a particular situation, and how those factors might be harmonized (Moreno 1995).

So, how do we decide better and worse? It is here that the revolutionary character of Dewey's view becomes evident. From a naturalistic perspective, we have to give up the idea that there is some fixed and finished realm of values (principles, goods, ends), established in advance and then brought into play to resolve conflicts among values in a particular situation. We have to replace the idea of moral deliberation as a *discovery* of preexisting moral principles and values, with the view of moral deliberation as a *transformative activity* by which experience is reconstructed via what Dewey called *dramatic imaginative rehearsal* of the consequences of pursuing certain values, principles, or dispositions in a specific situation. Moral deliberation is seen to be a form of aesthetic sensibility concerning the meaning of a situation and what is possible by way of ameliorative action. Just as scientific knowing results in the making of a new situation by instituting new meanings, discovering new relations, and realizing new values, likewise, moral deliberation resolves conflicts of values when it enriches the meaning of one's situation by exploring how experience might be changed under the influence of various values. The key is to deepen and enrich the meaning of one's situation, by expanding one's vision of what is possible and how it can be realized in a way that would reduce conflict of values: "Morals means growth of conduct in meaning, at least it means that kind of expansion in meaning which is consequent upon observations of the conditions and outcome of conduct. . . . In the largest sense of the word, morals is education. It is learning the meaning of what we are about and employing that meaning in action. The good, satisfaction, 'end,' or growth of present action in shades and scope of meaning is the only good within our control, and the only one, accordingly, for which responsibility exists" (Dewey 1922/1988, 194).

The radical character of this perspective can be seen in the metaphysics, and in the new conception of moral deliberation, that it entails. Dewey rejects the idea of a universe whose categories, moral and otherwise, are given in advance and precooked. If the moral universe is already baked and complete, then moral judgment would be nothing more than bringing cases under preestablished imperatives, calculating what will bring about the greatest good that is available, or cultivating virtues that are approved by society as productive of "the good."

Instead of a fixed, block moral universe, Dewey argues that we live in a universe that is in the making, where human intelligence can play a role in what transpires. If moral deliberation is imaginative problem solving, just like our other ways of inquiry involving need–search–satisfaction, then it starts with a universe in process, and it thus sees its mission as trying to be an intelligent force for directing change for the better within this evolving world: "Deliberation is actually an imaginative rehearsal of various courses of conduct. We give way, *in our mind*, to some impulse; we try, *in our mind*, some plan. Following its career through various steps, we find ourselves in imagination in the presence of the consequences that would follow; and as we then like and approve, or dislike and disapprove, these consequences, we find the original impulse or plan good or bad. Deliberation is dramatic and active, not mathematical and impersonal; and hence it has the intuitive, the direct factor in it" (Dewey & Tufts 1932/1989, 275; italics in original).

A good way to capture this distinction between discovering and creating or transforming moral values is Dewey's distinction between *conscience* and *conscientiousness*. According to the traditional notion of conscience, each person is born with an innate (god-given, reason-given, or evolutionarily developed) faculty that allows them (1) to discern what unconditional moral principles are binding on them and (2) to see how to apply the relevant principle to the present problematic case. In such a cosmos, what is good and right is given, and all of the moral principles we'll ever need are lodged right there in one's conscience, just waiting to be discovered, appreciated, and applied to life situations. Our duty is first to *discern* what's right, by consulting our conscience, and then to develop disciplined strength of will (i.e., willpower) to *do* what is right, what is morally required of us. The traditional view is that conscience is a moral faculty given at birth, activated by experience, and definitive of our moral obligations. Recently, this classic view of conscience has been replaced with the related notion of a "universal moral grammar" (Mikhail 2000, 2007), that John Mikhail defines as "an innate function or morality acquisition device that maps the child's early experience into the system of principles that constitutes the mature state of her moral competence (2008, 355).

In contrast, we are suggesting that there is no such thing as an innate moral faculty or moral grammar (Johnson 2014a). The only plausible notion of conscience is therefore just the unreflective intuitive operation of habits and values developed through past experience (individual and cultural) and thereby shaping, mostly unconsciously, our present valuing.

Instead of a completed and fixed universe, Dewey's metaphysics recognizes a universe in process, in which what is "given" to an individual at a point in time is merely the sedimented habits of valuings that constitute his or her present moral self. That "self as currently constituted" provides a *present, though temporary*, springboard for the temporally emerging self that is in the making through our deliberative activities. Our sedimented value prejudices have been habitually reinforced by past experience and entrenched cultural values, practices, and institutions. However, these norms may not be what is needed in response to new conditions that are generating value conflicts. Indeed, far from being the solution, sticking unreflectively to preestablished habits may actually be the problem, so the solution is to reconstruct or create new habits (and, thereby, a new moral self) more responsive to new, changing circumstances.

That reflective critical transformative capacity is what Dewey calls *conscientiousness*, which is a basic epistemic and moral virtue. It involves cultivating our capacities for criticism of value systems in the service of resolving value conflict: "The truly conscientious person not only uses a standard in judging, but is concerned to revise and improve this standard. . . . Only by thoughtfulness does one become sensitive to the far-reaching implications of an act; apart from continual reflection we are at best sensitive only to the value of special and limited ends" (Dewey & Tufts 1932/1989, 273). Conscientiousness is thus the virtue of developing and utilizing our cultivated capacities for criticism of value systems, as a basis for imaginatively rehearsing the experiential outcome of pursuing certain ends and values, and acting on certain principles, in order to decide which direction best resolves the conflicts at hand. And it bears reemphasizing that the "best" solution is not merely what an individual finds satisfactory, but requires the most comprehensive perspective possible, with a keen awareness of our own blinders to what may be most important in a given situation. Conscientiousness thus requires perceptiveness, empathic understanding of other people's experiences and values, emphasis on critical reflection, and openness to change and reconstructive activity.

It is for this reason that Philip Kitcher (2011) uses the term "the Ethical Project" to highlight the ongoing self-criticism, expansion of perspectives, and continual growth and refinement of meaning that characterize the most constructive moral deliberations. This, in turn, recalls Dewey's insistence that the (moral) self is always a *self-in-process*. The self as presently formed is an interpenetration of habits acquired from evolution and prior life experience. That present self shapes the way we identify moral problems and begin to investigate them, but that prior self gets reconstructed through the imaginative dramatic rehearsal of moral deliberation. The mark of moral rectitude is therefore not adherence to some absolute pregiven norms, but rather is to be judged by the growth of meaning acquired through the deliberative process.

Ethics naturalized is a "morality for humans" (Johnson 2014a). It disabuses us of any pretention to ultimates or absolutes. It humbles us, but at the same time gives us a realistic means of moral problem solving (Flanagan 1991, 2007). The results of such a process will never be final, though they may be the best we have for the moment. Present good is not the final endpoint of our reflections, but rather a temporary resting place from which to take on new challenges, in pursuit of what is better. That search for the better is a never-ending strenuous journey. It is hard to uncover and examine our deepest prejudices. It is hard to consider the experiences of all of those whose well-being is affected by your decisions and actions. It is hard to imagine creative and lasting resolutions of conflict. The conscientious inquirer will be prepared to reconsider prior deliberations in the face of new circumstances, which is one important indicator of moral growth.

Moral deliberation and judgments are not fundamentally different in character from aesthetic judgments. In his *Critique of Judgment* (1790/1987) Kant glimpsed this profound truth, in his convoluted argument that "beauty is the symbol of morality" (section 59), although he thought he had to pull back from placing aesthetic considerations at the heart of experience and thought, in order to preserve his account of the objectivity and universality of morals. It was Dewey who made aesthetics the cornerstone of his philosophy and of moral judgment (Alexander 1987, 2013; Johnson 2018). Acts of moral deliberation create new possibilities for meaningful action capable of changing our world, usually in some small measure, but occasionally with significant, far-reaching effects. As we saw in chapter 8, works of art creatively enact possibilities for meaning, revealing worlds we can inhabit via imaginative empathic engagement. In both aesthetic and moral explorations, notions of better and worse are decided by how well the envisioned possibilities resolve conflicts and liberate our capacities for intelligent action and future growth.

This conclusion is anxiety producing for those who have wanted to see moral judgments as objective and dispassionate, while denigrating aesthetic judgments as subjective and emotion based. Kant, for example, saw that there could not be rules for determining what is beautiful, but he failed to appreciate that this applies also to moral inquiry and judgments. So, he resorted to trying to derive moral laws from the concept of rational agency (i.e., a good will), while leaving aesthetic judgments in the limbo of demanding agreement, but without being able to ground that demand in any universal concepts. Dewey countered by pointing out the fundamentally qualitative character of both aesthetic and moral judgments. Moral deliberation is more like creating a work of art than it is a matter of either rule following or calculation of goodness. Both in artistic making and in the remaking of experience via moral inquiry, better and worse are objective considerations, insofar as some moral inquiries actually do enlarge our understanding, show us the consequences of certain value commitments, resolve conflicts, and harmonize our activities. Other judgments lead to confusion, conflict, narrowing of vision, impoverishment of meaning, and deadening or restricting of capabilities.

When push comes to shove, you cannot force another to agree with your preferred moral judgments. You can try to evoke from them empathic understanding, you can help them imagine what certain principles, values, and virtues might mean in different situations, and you can give your best account of a world operating under certain values, but we never, despite our wishes, had power to compel by dint of universal reason. Unless you can find some shared values, unless you and your interlocutor have goodwill toward one another, and unless you can offer a narrative of possibilities that makes sense for both of you, your prospects of achieving moral agreement are slim. We have to abandon notions of moral compulsion and see moral deliberation as aesthetically based, imaginative problem solving.

What May I Hope?

We come, at last, to the profound aspirational issue—what can I hope for in the sort of world defined by pragmatic naturalism? Most folks the world over hope their lives will be meaningful, in the sense of doing something that matters to themselves and to others. They want some of what they do to somehow make a positive contribution to humanity and/or to the world in which they live (plants, animals, the global environment, etc.).

Unfortunately, a modicum of reflection reminds us just how difficult "mattering" really is. When we are young and developing, our future seems to stretch out before us, as an indefinite expanse of indeterminate, alluring, and yet anxiety-inducing possibilities for living, valuing, and doing. Most of us are so caught up in school, work, and family life—with "getting on in life"—that we seldom pause to consider our ultimate fate. However, if we accept the best scientific understanding of our human condition, or even just what our experience tells us, we will realize that "nobody's getting out of here alive!" Albert Camus, in *The Myth of Sisyphus*, speaks of the "cruel mathematics that command our condition" (1955/1980, 69), by which he meant that 100 percent of us are going to die someday—some sooner, some later, but no one will escape our ultimate fate. For this and other reasons, that old grouch, Schopenhauer, warns us against rushing toward the future:

> In our early youth we sit before the life that lies ahead of us like children sitting before the curtain in a theater, in happy and tense anticipation of whatever is going to appear. Luckily we do not know what really will appear. For to him who does know, children can sometimes seem like innocent delinquents, sentenced

> not to death but to life, who have not yet discovered what their punishment will consist of. Nonetheless, everyone desires to achieve old age, that is to say a condition in which one can say: "Today it is bad, and day by day it will get worse—until at last the worst of all arrives." (Schopenhauer 1970, 8)

We have to face the issue of death head-on, without blinders. Each of us is going to die. There is no substantial corroborated evidence that humans have a disembodied soul capable of living on after the death of our body. On the contrary, there is overwhelming evidence that all our life-sustaining, cognitive, and emotional processes depend on a living, at least partially functioning, brain, operating a living, at least partly functioning, body, interacting with a material, interpersonal, and cultural environment. We have mountains of evidence concerning the conditions of our brain and body that sustain life, consciousness, and thought (Damasio 1999, 2010; Edelman & Tononi 2000; Moreno & Schulkin 2020). When our brain and central nervous system lose their ability to maintain basic life functions, it's "Lights Out" for us. Human life rarely exceeds a hundred years. To a teenager, a day in school may seem like an eternity, but to a seventy-year-old, every day seems to fly past.

What makes this a pressing existential issue for most people is the fact that mattering requires a living being to and for whom something matters. Camus understood this issue and its implications for our human condition. In the short novel *The Stranger*, the protagonist Meursault, sentenced to death for killing a man, is awaiting execution via the guillotine. He is repeatedly visited by the prison chaplain, who wants him to admit his guilt, confess his faith in Christ crucified, and ask God for forgiveness. In an explosion of pent-up rage and frustration, Meursault seizes the priest by his cassock and unleashes upon him a torrent of insults:

> Then, I don't know why, but something inside me snapped. I started yelling at the top of my lungs, and I insulted him and told him not to waste his prayers on me. . . . He seemed so certain about everything, didn't he? And yet none of his certainties was worth one hair of a woman's head. He wasn't even sure he was alive, because he was living like a dead man. Whereas it looked as if I was the one who'd come up empty handed. But I was sure about me, about everything, surer than he could ever be, sure of my life and sure of the death I had waiting for me. . . . And so? It was as if I had waited all this time for this moment and for the first light of this dawn to be vindicated. Nothing, nothing mattered, and I knew why. So did he. Throughout the whole absurd life I'd lived, a dark wind had been rising toward me from somewhere deep in my future, across years that were still to come, and as it passed, this wind leveled whatever was offered to me at the time. (Camus 1942/1988, 120)

From a naturalistic perspective, the same "dark wind" of death is blowing toward each and every one of us. Nobody gets out alive, and so the big question arises: Does this death wind level everything in its path, which is to say, rob everything of any lasting meaning? Meursault says yes, "Nothing, nothing mattered," not even life and death. In this novel, Camus seems to be agreeing with Meursault, who is ultimately more on trial for challenging his culture's values and beliefs than for murdering an Arab. There is no ultimate redemption at the end. Meursault awakens from a cleansing slumber following his tirade toward the chaplain, only to feel open to "the gentle indifference of the world. Finding it so much like myself—so like a brother, really—I felt that I had been happy and that I was happy again. For everything to be consummated, for me to feel less alone, I had only to wish that there be a large crowd of spectators the day of my execution, and that they greet me with cries of hate" (Camus 1942/1988, 123).

This is a bleak picture of the human condition. The vehemence of the huge crowd pouring out their rage on Meursault at his execution would be vindication for Meursault, because such a rage would be indicative of how much they hate what he stands for, how much they feel threatened by his values, and how much they fear that he may be right, after all. Meursault (his life, his perspective, his very existence) is an affront to his fellow citizens, because he pulls back the veil of bad faith to reveal the absence of any supposed world beyond this world of our bodily habitation. In a similar vein, Nietzsche (1895/1990) had earlier argued that the values that people claim are rooted in the supernatural are, in fact, nihilistic values, insofar as they are predicated on a nonexistent transcendent realm which denigrates our "fallen" human bodily condition here in the natural world. The Nietzsche who was so widely accused by his society of being a nihilist, thus turns the tables on them by charging that *their* otherworldly values are nihilistic, insofar as they deny this very world of nature which we *all* inhabit, and from which our values actually arise.

Can any of us live comfortably, or at least with integrity, in a world in which "nothing matters?" Probably not. Even Camus, the great chronicler of what Heidegger (1927/1962) called our "being-toward-death," went on to write *The Plague* (1948), in which the doctor, who has a chance to flee a plague-ridden walled city, chooses, out of solidarity with his fellow sufferers and his caring concern for them, to remain in the city to help others, thereby facing almost certain death. The closed off walled city represents our world, a world in which the dark wind of death eventually levels us all, and from which there is no escape. Camus seems to admire the doctor for his honest recognition of our human condition, while still acting to make the world a better place, even without any ultimate justification or hope.

However, if Meursault was right that "nothing matters," then why should the doctor's actions be seen to matter, to be admirable, to have value? Camus thus seems to realize that we need for *something* to matter, even if it can never be a lasting value. Those moments that matter provide great sustenance for fundamental meaning in our lives. Camus's implicit respect for the doctor who remains behind to nurse the sick and dying, calls to mind Schopenhauer's claim that the proper response to our absurd condition, where our highest capacities are the source of our greatest suffering, is to dwell with others as companions in misery, fellow sufferers needing care and compassion. When Camus was young he aspired to a career as a football (soccer) player, but an injury ended that hope. Camus once said in an interview that his morality was based primarily on the morality of sporting teams, which value fraternity, team spirit, and common purpose. The idea that we're in this life together, staying loyal to one another, and working together toward shared goals, may give us a sense of purpose and mattering, but it cannot cancel the fact that each of us eventually dies our own death.

So, what may we hope? In addressing this fundament question, Dewey quotes at great length an extraordinarily poetic and beautiful passage from Oliver Wendell Holmes, about what our realistic human aspirations amount to. Holmes begins with our intrinsic situatedness in ongoing natural processes, from which, and in which, our values, ideals, and hopes arise. We are thrown into nature and have to muddle about in search of some way of living and finding meaning. When we realize what an insignificant part of the universe we are, and when we contemplate our final encounter with the dark wind, is there any justification for continuing to struggle, for taking ameliorative action? Justice Holmes says yes:

> That the universe has in it more than we understand, that the private soldiers have not been told the plan of campaign, or even that there is one . . . has no bearing on our conduct. We still shall fight—all of us because we want to live, some, at least, because we want to realize our spontaneity and prove our powers, for the joy of it, and we may leave to the unknown the supposed final valuation of that which in any event has value for us. It is enough for us that the universe has produced us and has within it, as less than it, all that we believe and love. (Quoted in Dewey 1925/1981, 313)

This strategy for living has its place in a naturalistic perspective. Lacking an understanding of any overarching plan or purposiveness for existence, yet we persist and struggle, just because there are things we care about in our lives. Those things about which an individual might care do not have to come from god as the ground of being. Nor do they require any universal rational justification. What matters is that we happen to care about certain things, whatever they might be, because of our physical needs, temperament, upbringing, or idiosyncratic experiences. We just do care, and what we care about matters, not absolutely, but *to us, while we live.*

From a naturalistic point of view, there is no transcendence toward a supernatural realm of existence, meaning, purpose, or value, but only the immanent transcendence of being part of an ongoing process to which we contribute. In Holmes's lovely words:

> If we think of our existence not as that of a little god outside, but as that of a ganglion within, we have the infinite behind us. It gives us our only but our adequate significance. If our imagination is strong enough to accept the vision of ourselves as parts inseparable from the rest, and to extend our final interest beyond the boundary of our skins, it justifies even the sacrifice of our lives for ends outside of ourselves. The motive to be sure is the common wants and ideals that we find in man. Philosophy does not furnish motives, but it shows men that they are not fools for doing what they already want to do. (Quoted in Dewey 1925/1981, 313)

What keeps us from being fools, as we strive and strut our precious hours on the stage of life? Holmes's romantic side emerges here. We are what John Donne called "a piece of the continent, a part of the main" (1624/1839, 574–575). We exist in solidarity with some of our fellows. We have only the most modest capacities for improving our lot and making the world better, but, says Holmes, we are not self-deceived in believing that amelioration is a genuine possibility. All of this, of course, is predicated on the possibility of our finding activities that matter to us. Without meaningful ends-in-view, we have no motivational impetus to endeavor and achievement.

From a pragmatic naturalistic perspective, there is no access to eternity. *Everything* is finite and contingent, and there are no transcendent, absolute values. Therefore, saying, as Holmes does, that we are part of something larger than ourselves cannot be a miracle cure for our finiteness. It may bring some comfort to feel connected to the ever-developing nature of which we are a part, or to feel solidarity with our fellow travelers through life, but there is no ultimate justification of any absolute value or meaning.

Whatever meaning we have or make is contingent. It thus depends on the maintenance of certain motivational processes within us and the availability of satisfactions that we find interesting and alluring. Contrary to what William James (1890) proposed as a way to deal with his life-threatening depression, you cannot *will* yourself to find something meaningful or something that matters. One of the sad things about our human condition is that genetic influences, illnesses, trauma, and brain lesions can rob us of the very capacities that allow us to experience something as meaningful, or to find something that matters. We no doubt hope that medical advances may reduce some of these assaults on our sense of purposiveness, meaning, and mattering, but we are not yet there. There is thus an element of grace upon which the possibility of a meaningful existence rests. This is not the undeserved gift of the grace of God that will supposedly save you from sin, death, injustice, meaninglessness, and aloneness. It is rather a natural grace, where natural and social processes sometimes merge to generate in you a feeling of connectedness, a sense of having meaningful projects, and an awareness that some things actually do matter to you—do give you something or someone that you care about. It is a grace bestowed upon you partly by others who have gone before you and left you their values for your critical assessment and appropriation: "The things in civilization we most prize are not of ourselves. They exist by grace of the doings and sufferings of the continuous human community in which we are a link. Ours is the responsibility of conserving, transmitting, rectifying and expanding the heritage of values we have received that those who come after us may receive it more solid and secure, more widely accessible and more generously shared than we have received it. Here are all the elements for a religious faith that shall not be confined to sect, class, or race" (Dewey 1934b/1986, 87).

This sense of the possibilities for meaning, nurturant values, and intelligent inquiry is Dewey's religion. He asks us to replace *religions*—characterized by supernatural metaphysics, catalogues of supposedly absolute values and principles, institutional practices, and the promise of ultimate enlightenment and salvation—with an appreciation of the *religious quality* of ordinary, yet meaningful daily experience. This religious quality is our sense of the possibilities for meaningful activities, based on a faith in human intelligence to help us deal with the slings and arrows of outrageous fortune that we encounter in our lives: "Any activity pursued in behalf of an ideal end against obstacles and in spite of threats of personal loss because of conviction of its general and enduring value is religious in quality. Many a person, inquirer, artist, philanthropist, citizen, men and women in the humblest walks of life, have achieved, without presumption and with display, such unification of themselves and of their relations to the conditions of their existence" (Dewey 1934b/1986, 27).

Dewey had a very modest and humble sense of himself, his powers, and his place in nature. He waxes eloquent with his own Romantic justification of human striving, when he reminds us that, so long as we live, nature will present us with challenges for our powers, meanings to be discovered, and possibilities for values we can realize:

> Fidelity to the nature to which we belong, as parts however weak, demands that we cherish our desires and ideals till we have converted them into intelligence, revised them in terms of the ways and means which nature makes possible. When we have used our thought to its utmost and have thrown into the moving unbalanced balance of things our puny strength, we know that though the universe slay us still we may trust, for our lot is one with whatever is good in existence. As far as we are concerned it is the only condition for it alone is in our power. To ask more than this is childish; but to ask less is a recreance no less egotistic. (Dewey 1925/1981, 314)

Savor for the moment Dewey's lovely phrase, "we have thrown into the moving unbalanced balance of things our puny strength." We struggle in a precarious world that continually offers us challenges and frustrates our goals. There is an "unbalanced balance," insofar as, in our ongoing striving which is always confronted with new challenges, there is nevertheless an overarching dynamic balance of allostatic adjustment that carries us forward and gives us hope for at least some measure of success with our diverse projects. Absent that reinforcement, we would be lost in despair, thrown back into Meursault's world in which "nothing matters."

In our endeavors to find meaningful activities and to help make things better, we are not alone. We are part of what Dewey called the "*community of life*" (1922/1988, 226), in which we together live, suffer, and strive: "Within the flickering inconsequential acts of separate selves dwells a sense of the whole which claims and dignifies them. In its presence we put off mortality and live in the universal. The life of the community in which we live and have our being is the fit symbol of this relationship" (Dewey 1922/1988, 227).

The world as naturalism sees it, has no ultimate purpose, no supernatural basis, no promise of an afterlife, and no remedy for death. However, as

Alexander (2013) suggests, our world is the site of our meaningful strivings, our human eros for meaning. We find that meaning where we find it, in things, people, and causes we find ourselves caring about. We can use our puny strength to help make the world better, rather than worse. Nobody—not Jesus, or Moses, or Mohammed, or the Buddha, or the Beatles—can save us from our human condition, but in and through them, some of us may find meaningful ways to carry on and to throw in with the better, rather than the worse. If we find meaning, it will be via the ways we have found projects and commitments that matter to us, driving us on to better things.

Appendix

Richard Rorty and Richard Bernstein, both students of Paul Weiss, John Smith at Yale, and John McDermott at Queens would carry the banner for pragmatism, introducing it to a whole new group of thinkers across the globe. However, neither emphasized the central importance of notions like experience and nature so essential to the thinking of the classical pragmatists, nor did they appreciate the integration of the sciences into philosophy.

This rediscovery of pragmatism began with the revival of the classic founders of the movement: Charles Sanders Peirce, William James, John Dewey, Oliver Wendell Holmes, Jr., and George Herbert Mead, but this short and parochial list of canonical figures soon expanded to embrace more socially and politically engaged thinkers, such as Jane Addams (1902), Alain Locke (1924), W. E. B. du Bois (1903), and others. By the 1990s, this diversification blossomed into an explosion of interest in pragmatist views from all directions. With apologies to all those that we have left out, here are a few representative works that contributed to this opening up of new perspectives, in fields such as feminism (Marilyn Fischer 2000; Judith Green 2009; Lisa Heldke 2019; Charlene Haddock Seigfried 1996; Shannon Sullivan 2001), ethics (Steven Fesmire 2003, 2015; Mark Johnson 1993, 2014a, 2014b; Philip Kitcher 2011; Jonathan Moreno 1999; Gregory Pappas 2008), Native American philosophy (Patricia Hill Collins 2011; Vine Deloria, Jr. 1999; Thomas Norton-Smith 2010; Scott Pratt 2002; Laurelyn Whitt 2009; Bruce Wilshire 2000), race (Leonard Harris 2021; Lucius Outlaw 1996; Shannon Sullivan 2001; Cornel West 1989), metaphysics and epistemology (Richard Bernstein 1983; David Hildebrand 2003; Christopher Hookway 1984; Hilary Putnam 1983; Ralph Sleeper 1986), environmental (Andrew Light 1996; Mark Sagoff 2000; Paul Thompson & Thomas Hilde 2000), technology (Larry Hickman 2001, 2019), Latin American philosophy (José Medina 2011; José-Antonio Orosco 2011; Gregory Pappas 2011), aesthetics (Thomas Alexander 1987, 2013; Mark Johnson 2007, 2018; Richard Shusterman 2000, 2008), religion (Robert Neville 1992; John Smith 1961; Cornel West 1989; Bruce Wilshire 2000), comparative philosophy (David Hall & Roger Ames 1995; Robert Neville 1992), education (Nel Noddings 2019; Andrea English & Christine Doddington 2019), genealogical pragmatism (Colin Koopman 2009, 2013; Joseph Margolis 2002; John Stuhr 1997, 2003), biological and physical sciences (Peter Godfrey-Smith 1998; Philip Kitcher 2012, 2018; Hillary Putnam 1981), cognitive science (Shaun Gallagher 2016; Roman Madzia 2016; Teed Rockwell 2005; Pierre Steiner 2014; Italio Testa & Fausto Caruana 2020), phenomenology (Anthony Chemero 2009; Shaun Gallagher 2018; Sandra Rosenthal & Patrick Bourgeois 1980), political and social theory (Richard Bernstein 1983; Jurgen Habermas 1967; Richard Rorty 1982, 1990; John Stuhr 2016; Robert Talisse 2009), law (Brian Butler 2017; Benjamin Cardoza 1928; Tom Grey 1989; Fred Kellogg 1992; Richard Posner 1993; Steven Winter 2001, 2021), neuroscience (Fausto Caruana & Italo Testa 2021; Mark Johnson 2007, 2018; Jay Schulkin 2009, 2015; Tibor Solymosi & John Shook 2014), analytic pragmatism (Robert Brandom 2008, 2011; Cheryl Misak 2016; Huw Price 2013).

References

Addams, J. 1902. *Democracy and Social Ethics*. New York: Macmillan.

Adolphs, R. 1999. Social cognition and the human brain. *Trends in Cognitive Sciences* 3 (12): 469–479.

Adolphs, R., Denburg, N. L., & Tranel, D. 2001. The amygdala's role in long-term declarative memory for gist and detail. *Behavioral Neuroscience* 115: 983–992.

Alexander, T. M. 1987. *John Dewey's Theory of Art, Experience, and Nature: The Horizons of Feeling*. Albany, NY: SUNY Press.

Alexander, T. M. 2013. *The Human Eros: Eco-Ontology and the Aesthetics of Existence*. New York: Fordham University Press.

Arendt, H. 1958. *The Human Condition*. Chicago: University of Chicago Press.

Aristotle. 1941. *Poetics*. Translated by I. Bywater. In R. McKeon (Ed.), *The Basic Works of Aristotle*. New York: Random House.

Ayer, A. J. 1936. *Language, Truth, and Logic*. London: Gollancz.

Baron, J. 1988. *Thinking and Deciding*. Cambridge: Cambridge University Press.

Barrett, L. F., & Simmons, W. K. 2015. Interoceptive predictions in the brain. *Nature Reviews Neuroscience* 16 (7): 419–429.

Barsalou, L. 1999. Perceptual symbol systems. *Behavioral and Brain Sciences* 22: 577–660.

Barthes, R. 1968. *Elements of Semiology*. Translated by A. Lavers and C. Smith. New York: Hill & Wang.

Bell, C. 1913. *Art*. New York: Frederick A. Stokes.

Bergen, B. 2012. *Louder Than Words: The New Science of How the Mind Makes Meaning*. New York: Basic Books.

Bergmann, G. 1960. Critical review of P.F. Strawson. Individuals: An essay in descriptive metaphysics. *Journal of Philosophy* 57 (19): 601–922.

Bergson, H. 1946. *The Creative Mind: An Introduction to Metaphysics.* New York: Philosophical Library.

Berlyne, D. E. 1954. A theory of human curiosity. *British Journal of Psychology* 45 (3): 180.

Bernard, C. 1865/1957. *An Introduction to the Study of Experimental Medicine.* New York: Dover Publications.

Bernstein, R. 1983. *Beyond Subjectivism and Relativism.* Philadelphia: University of Pennsylvania Press.

Berridge, K. C. 2004. Motivation concepts in behavioral neuroscience. *Physiology and Behavior* 81 (1–2): 179–209.

Berridge, K. C. 2006. The debate over dopamine's role in reward: The case for incentive salience. *Psychopharmacology* 191: 391–431.

Berthoz, A. 2002. *The Brain's Sense of Movement.* Cambridge, MA: Harvard University Press.

Block, N., Flanagan, O., & Guzeldere, G. 1999. *The Nature of Consciousness.* Cambridge, MA: MIT Press.

Brandom, R. 2008. *Between Saying and Doing: Toward an Analytic Pragmatism.* Oxford: Oxford University Press.

Brandom, R. 2011. *Perspectives on Pragmatism: Classical, Recent, and Contemporary.* Cambridge, MA: Harvard University Press.

Brentano, F. 1874/1973. *Psychology from an Empirical Standpoint.* Translated by A. Rancurello, D. Terrill, & L. McAlister. London: Routledge.

Bullough, E. 1912. "Psychical distance" as a factor in art and an aesthetic principle. *British Journal of Psychology* 5 (2): 87–118.

Butler, B. 2017. *The Democratic Constitution: Experimentalism and Interpretation.* Chicago: University of Chicago Press.

Buzsaki, G. 2006. *Rhythms of the Brain.* Oxford: Oxford University Press.

Buzsaki, G., Peyrache, A., & Kubie, J. 2014. Emergence of cognition from action. *Cold Springs Harbor Laboratory Press* 23: 1–15.

Cabanac, M. 1979. Sensory pleasure. *Quarterly Review of Biology* 54 (3): 1–29.

Cadwallader, T. C. 1975. Peirce as an experimental psychologist. *Transactions of the C.S. Peirce Society* 11: 167–186.

Calvert, G. A., Bullmore, E. T., Brammer, M. J., Campbell, R., Williams, S. C. R., McQuire, P. K., Woodruff, P. W. R., Iverson, S. D., & David, A. S. 1997. Activation of auditory cortex during silent lip-reading. *Science* 276: 593–596.

Camus, A. 1942/1988. *The Stranger.* Translated by M. Ward. New York: Random House.

Camus, A. 1948. *The Plague.* London: Hamish Hamilton.

Camus, A. 1955/1980. The myth of Sisyphus. In D. Cheney & S. Saunders (Eds.), *The Meaning of Life: Questions, Answers and Analysis* (pp. 65–75). Englewood Cliffs, NJ: Prentice-Hall.

Cannon, W. B. 1915, 1929. *Bodily Changes in Pain, Hunger, Fear and Rage.* New York: Harper Torchbooks.

Cannon, W. B. 1932/1963. *The Wisdom of the Body.* New York: W. W. Norton.

Cardoza, B. 1928. *The Paradoxes of Legal Science.* New York: Columbia University Press.

Carnap, R. 1928/1967. *The Logical Structure of the World: Pseudoproblems in Philosophy.* Berkeley: University of California Press.

Carter, C. S. 2014. Oxytocin pathways and the evolution of human behavior. *Annual Review of Psychology* 65 (1): 17–39.

Cartwright, N. 1983. *How the Laws of Physics Lie.* Oxford: Clarendon Press.

Caruana, F., & Testa, I. (Eds.) 2021. *Habits: Pragmatist Approaches from Cognitive Science, Neuroscience, and Social Theory.* Cambridge: Cambridge University Press.

Cassirer, E. 1929/1953. *The Philosophy of Symbolic Forms* (Vol. 3). New Haven, CT: Yale University Press.

Cavell, S. 1998. *In Quest of the Ordinary: Lines of Skepticism and Romanticism.* Chicago: University of Chicago Press.

Chalmers, D. 1995. Facing up to the problem of consciousness. *Journal of Consciousness Studies* 2 (3): 200–219.

Chao, L. L., Weisberg, J., & Martin, A. 2002. Experience-dependent modulation or category-related cortical activity. *Cerebral Cortex* 12: 545–551.

Chemero, A. 2009. *Radical Embodied Cognitive Science.* New York: Bradford.

Chemero, A., & Kaufer, S. 2016. Pragmatism, phenomenology, and extended cognition. In R. Madzia & M. Jung (Eds.), *Pragmatism and Embodied Cognitive Science* (pp. 57–72). Berlin: Walter de Gruyter.

Cheney, D. L., & Seyfarth, R. M. 1990. *How Monkeys See the World.* Chicago: University of Chicago Press.

Cheney, D. L., & Seyfarth, R. M. 2007. *Baboon Metaphysics.* Chicago: University of Chicago Press.

Churchland, P. 1986. *Neurophilosophy: Toward a Unified Science of the Mind-Brain.* Cambridge, MA: MIT Press.

Clark, A. 1998. *Being There: Bringing Brain, Body, and World Together*. Cambridge, MA: Bradford Books; MIT Press.

Clark, A. E. 2016. *Surfing Uncertainty: Prediction, Action, and the Embodied Mind*. Oxford: Oxford University Press.

Clark, A., & Chalmers, E. 1998. The extended mind. *Analysis* 58 (1): 7–19.

Coegnarts, M., & Kravanja, P. (Eds.) 2015. *Embodied Cognition and Cinema*. Leuven: Leuven University Press.

Cohen, M., & Dennett, C. 2011. Consciousness cannot be separated from function. *Trends in Cognitive Science* 8 (15): 358–364.

Collins, P. H. 2011. Piecing together a genealogical puzzle: Intersectionality and American pragmatism. *European Journal of Pragmatism and American Philosophy* 3 (2): 88–112.

Craig, W. 1918. Appetites and aversions as constituents of instinct. *Biological Bulletin* 34: 91–107.

Crews, D. 2011. Epigenetic modifications of brain and behavior: Theory and practice. *Hormones and Behavior* 59 (3): 393–398.

Crick, F., & Koch, C. 1990. Towards a neurobiological theory of consciousness. *Seminars in the Neurosciences* 2: 263–275.

Crippen, M., & Schulkin, J. 2020. *Mind Ecologies: Body, Brain, and World*. New York: Columbia University Press.

Damasio, A. 1994. *Descartes' Error: Emotion, Reason, and the Human Brain*. New York: G. P. Putnam's Sons.

Damasio, A. 1996. The somatic marker hypothesis and the possible functions of the prefrontal cortex. *Philosophical Transactions of the Royal Society of London* 351 (1346): 1413–1420.

Damasio, A. 1999. *The Feeling of What Happens: Body and Emotion in the Making of Consciousness*. New York: Harcourt Brace.

Damasio, A. 2003. *Looking for Spinoza: Joy, Sorrow, and the Feeling Brain*. New York: Basic Books.

Damasio, A. 2010. *Self Comes to Mind: Constructing the Conscious Brain*. New York: Pantheon Books.

Damasio, A. 2018. *The Strange Order of Things: Life, Feeling, and the Making of Cultures*. New York: Pantheon Books.

Danto, A. 1964. The artworld. *Journal of Philosophy* 61 (19): 71–84.

Darwin, C. 1859/1958. *On The Origin of Species*. New York: Mentor Book.

Darwin, C. 1871/1874. *Descent of Man*. New York: Rand McNally & Company.

Darwin, C. 1872/1998. *The Expression of the Emotions in Man and Animals*. Oxford: Oxford University Press.

Deloria, V., Jr. 1999. *Spirit and Reason*. Wheat Ridge, CO: Fulcrum Publishing.

Dennett, D. 1984. *Elbow Room*. Cambridge, MA: MIT Press.

Dennett, D. 1991. *Consciousness Explained*. New York: Little Brown.

Denton, D. 1982. *The Hunger for Salt*. Berlin: Springer-Verlag.

Denton, D. 2005. *The Primordial Emotions*. Oxford: Oxford University Press.

Derrida, J. 1967/1976. *Of Grammatology*. Translated by G. Spivak. Baltimore, MD: Johns Hopkins University Press.

Desimone, R. 1996. Neural mechanisms for visual memory and their role in attention. *Proceedings of the National Academy of Sciences* 93 (24): 13494–13499.

Dethier, V. J. 1968. Chemosensory input and taste discrimination in the blowfly. *Science* 161 (3839): 389–391.

Dewey, J. 1896. The reflex arc concept in psychology. *Psychological Review* 3: 357–370.

Dewey, J. 1905. The postulate of immediate empiricism. *Journal of Philosophy, Psychology, and Scientific Methods* 2: 333–339.

Dewey, J. 1910. The influence of Darwin on philosophy. In Vol. 4 of *The Middle Works, 1899–1924* (pp. 3–14). Edited by J. A. Boydston. Carbondale: Southern Illinois University Press.

Dewey, J. 1922/1988. *Human Nature and Conduct*. Vol. 14 of *The Middle Works, 1899–1924*. Edited by J. A. Boydston. Carbondale: Southern Illinois University Press.

Dewey, J. 1925/1981. *Experience and Nature*. Vol. 1 of *The Later Works, 1925–1953*. Edited by J. A. Boydston. Carbondale: Southern Illinois University Press.

Dewey, J. 1929/1984. *The Quest for Certainty*. Vol. 4 of *The Later Works, 1925–1953*. Edited by J. A. Boydston. Carbondale: Southern Illinois University Press.

Dewey, J. 1930a/1988. Qualitative thought. In Vol. 5 of *The Later Works, 1925–1953* (pp. 243–262). Edited by J. A. Boydston. Carbondale: Southern Illinois University Press.

Dewey, J. 1930b/1988. Three independent factors in morals. In Vol. 5 of *The Later Works, 1925–1953* (pp. 279–288). Edited by J. A. Boydston. Carbondale: Southern Illinois University Press.

Dewey, J. 1934a/1987. *Art as Experience*. Vol. 10 of *The Later Works, 1925–1953*. Edited by J. A. Boydston. Carbondale: Southern Illinois University Press.

Dewey, J. 1934b/1986. *A Common Faith*. Vol. 9 of *The Later Works, 1925–1953*. Edited by J. A. Boydston. Carbondale: Southern Illinois University Press.

Dewey, J. 1938/1991. *Logic: The Theory of Inquiry*. Vol. 12 of *The Later Works, 1925–1953*. Edited by J. A. Boydston. Carbondale: Southern Illinois University Press.

Dewey, J. 1939/1991. *Theory of Valuation*. Vol. 13 of *The Later Works, 1925–1953*. Edited by J. A. Boydston. Carbondale: Southern Illinois University Press.

Dewey, J., & Tufts, J. H. 1932/1989. *Ethics*. Vol. 7 of *The Later Works, 1925–1953*. Edited by J. A. Boydston. Carbondale: Southern Illinois University Press.

Dickie, G. 1974. *Art and the Aesthetic: An Institutional Analysis*. Ithaca, NY: Cornell University Press.

Dickinson, A. 1980. *Contemporary Animal Learning*. Cambridge: Cambridge University Press.

Di Paolo, E. 2018. The enactive conception of life. In A. Newen, L. De Druin, & S. Gallagher (Eds.), *The Oxford Handbook of 4-E Cognition* (pp. 71–94). Oxford: Oxford University Press.

Di Paolo, E., Cuffari, E., & De Jaegher, H. 2018. *Linguistic Bodies: The Continuity between Life and Language*. Cambridge, MA: MIT Press.

Dobzhansky, T. C. 1962. *Mankind Evolving*. New Haven, CT: Yale University Press.

Donagan, A. 1977. *The Theory of Morality*. Chicago: University of Chicago Press.

Donald, M. 1991. *Origins of the Modern Mind*. Cambridge, MA: Harvard University Press.

Donald, M. 2001. *A Mind So Rare: The Evolution of Human Consciousness*. New York: Norton.

Donald, M. 2004. Hominid enculturation and cognitive evolution. In J. M. Luraciello, J. A. Hudson, R. Fivush, & P. J. Bauer (Eds.), *The Development of the Mediated Mind*. Mahwah, NJ: Erlbaum Press.

Donne, J. 1624/1839. No man is an island. *Devotions on Emergent Occasion*, Meditation XVI. *In H. Alford (Ed.), The Works of John Donne* (pp. 574–575). London: John W. Parker.

Du Bois, W. E. B. 1903. *The Souls of Black Folk*. Chicago: A. C. McClurg.

Dunbar, R. I. M. 1998. The social brain hypothesis. *Evolutionary Anthropology* 6: 178–190.

Dunbar, R. I. M., & Shultz, S. 2007. Evolution in the social brain. *Science* 317: 1344–1347.

Eagleman, D. 2011. *Incognito: The Secret Lives of the Brain*. New York: Random House.

Edelman, G. 2004. *Wider than the Sky: The Phenomenal Gift of Consciousness*. New Haven, CT: Yale University Press.

Edelman, G., & Tononi, G. 2000. *A Universe of Consciousness: How Matter Becomes Imagination*. New York: Basic Books.

Emery, N. J. 2000. The eyes have it: The neuroethology, function and evolution of social gaze. *Neuroscience and Biobehavioral Reviews* 24: 581–604.

Engel, A. K., Friston, K. J., & Kragic, D. 2015. *The Pragmatic Turn: Toward Action-Oriented Views in Cognitive Science*. Cambridge, MA: MIT Press.

English, A., & Doddington, C. 2019. Dewey, aesthetic experience, and education for humanity. In S. Fesmire (Ed.), *The Oxford Handbook of Dewey* (pp. 411–444). Oxford: Oxford University Press.

Fauconnier, G., & Turner, M. 2002. *The Way We Think: Conceptual Blending and the Mind's Hidden Capacities*. New York: Basic Books.

Feldman, J. 2006. *From Molecule to Metaphor: A Neural Theory of Language*. Cambridge, MA: MIT Press.

Fesmire, S. 2003. *John Dewey and Moral Imagination*. Bloomington: University of Illinois Press.

Fesmire, S. 2015. *Dewey*. New York: Routledge.

Fichte, J. G. 1796/2000. *Foundations of Natural Right*. Translated by M. Baur. Cambridge: Cambridge University Press.

Fisch, M. H. 1947. Evolution in American philosophy. *Philosophical Review* 5 (4): 357–373.

Fischer, M. 2000. Jane Addams's feminist ethics. In C. T. Tougas & S. Ebenreck (Eds.), *Presenting Women Philosophers*. Philadelphia: Temple University Press: 51–58.

Fitzsimons, J. T. 1998. Angiotensin, thirst and sodium appetite. *Physiological Reviews* 78 (3): 583–687.

Flanagan, O. 1991. *Varieties of Moral Personality: Ethics and Psychological Realism*. Cambridge, MA: Harvard University Press.

Flanagan, O. 1992. *Consciousness Reconsidered*. Cambridge, MA: MIT Press.

Flanagan. O. 2007. *The Really Hard Problem: Meaning in a Material World*. Cambridge, MA: MIT Press.

Fodor, J. A. 1976. *The Language of Thought*. Cambridge, MA: Harvard University Press.

Friston, K. 2010. The free-energy principle: A unified brain theory? *Nature Reviews Neuroscience* 11 (2): 127–138.

Friston, K., FitzGerald, T., Rigoli, F., Schwartenbeck, P., O'Doherty, J., & Pezzulo, G. 2016. Active inference and learning. *Neuroscience and Biobehavioral Reviews* 68: 862–879.

Frith, C. D. 2008. Social cognition. *Philosophical Transactions of the Royal Society B: Biological Sciences* 363 (1499): 2033–2039.

Gadamer, H. G. 1975. *Truth and Method.* Translated by W. Glen-Doepel. New York: Crossroads.

Gallagher, S. 2005. *How the Body Shapes the Mind.* Oxford: Oxford University Press.

Gallagher, S. 2016. Pragmatic interventions into enactive and extended conceptions of cognition. In R. Madzia & M. Jung (Eds.), *Pragmatism and Embodied Cognitive Science* (pp. 17–34). Berlin: Walter de Gruyter.

Gallagher, S. 2018. Rethinking nature: Phenomenology and a nonreductionist cognitive science. *Australasian Philosophical Review* 2 (2): 125–137.

Gallistel, C. R. 1982. *The Organization of Action: A New Synthesis.* Hillsdale, NJ: Lawrence Erlbaum.

Gallistel, C. R. 1993. *The Organization of Learning.* Cambridge, MA: MIT Press.

Garcia, J., Hankins, W. G., & Rusiniak, K. W. 1974. Behavioral regulation of the milieu interne in man and rat. *Science* 185: 824–831.

Gendlin, E. 1991. Crossing and dipping: Some terms for approaching the interface between natural understanding and logical formulation. *Minds and Machines* 5 (4): 547–560.

Gibson, J. J. 1979. *The Ecological Approach to Visual Perception.* Boston: Houghton Mifflin.

Gigerenzer, G. 2000. *Adaptive Thinking.* Oxford: Oxford University Press.

Gigerenzer, G. 2007. *Gut Feelings.* New York: Viking Press.

Gilligan, C. 1982. *In a Different Voice.* Cambridge, MA: Harvard University Press.

Godfrey-Smith, P. 1998. *Complexity and Function of Mind in Nature.* Cambridge: Cambridge University Press.

Goodman, N. 1978. *Ways of Worldmaking.* Indianapolis: Hackett.

Gopnik, A., & Meltzoff, A. 1997. *Words, Thoughts and Theories.* Cambridge, MA: MIT Press.

Gould, S. J. 2002. *The Structure of Evolutionary Theory.* Cambridge, MA: Harvard University Press.

Gould, S. J., & Eldridge, N. 1977. Punctuated equilibria: The tempo and mode of evolution reconsidered. *Paleobiology* 3: 115–151.

Graybiel, A. M. 1998. The basal ganglia and chunking of action repertoires. *Neurobiology of Learning & Memory* 70 (1-2): 119–136.

Green, J. M. 2009. *Pragmatism and Social Hope*. New York: Columbia University Press.

Greene, J. 2014. Beyond point and shoot morality: Why cognitive neuroscience matters for ethics. *Ethics* 124: 695–726.

Greene, J., Sommerville, R., Nystrom, L. E., Darley, J., & Cohen, J. 2001. An fMRI investigation of emotional engagement in moral judgment. *Science 293 (5537): 2105–2108.*

Gregory, M. 2009. *Shaped by Stories: The Ethical Power of Narrative*. Notre Dame, IN: University of Notre Dame Press.

Gregory, R. L. 1966. *Eye and Brain*. New York: McGraw-Hill.

Grey, T. 1989. Holmes and legal pragmatism. *Stanford Law Review* 41 (4): 787–870.

Griffin, D. 1974. *Listening in the Dark*. New York: Dover Press.

Gross, C. G. 1999. *Brain, Vision and Memory*. Cambridge, MA: MIT Press.

Habermas, J. 1967. *On the Logic of the Social Sciences*. Cambridge: Cambridge University Press.

Hacking, I. 1999. *The Taming of Chance*. Cambridge: Cambridge University Press.

Haidt, J. 2012. *The Righteous Mind: Why Good People Are Divided by Politics and Religion*. New York: Pantheon Books.

Hall, D., & Ames, R. 1995. *Anticipating China: Thinking through the Narrative of Chinese and Western Culture*. Albany, NY: SUNY Press.

Hanson, N. 1958. *Patterns of Discovery: An Inquiry into the Conceptual Foundations of Science*. Cambridge: Cambridge University Press.

Harris, L. 2021. *Philosophy Born of Struggle: Afro-American Philosophy from 1917*. University of Iowa: Kendall Hunt.

Haxby, J. V., Gobbini, M. I., Furey, M. L., Ishai, A., Schouten, J. L., & Pietrini, P. 2001. Distributed and overlapping representations of faces and objects in ventral temporal cortex. *Science* 293 (5539): 2425–2430.

Hebb, D. O. 1949. *The Organization of Behavior*. New York: Wiley.

Heelan, P. A., & Schulkin, J. 1998. Hermeneutical philosophy and pragmatism: A philosophy of the sciences. *Synthese* 115 (3): 269–302.

Hegel, G. W. F. 1807/2018. *Phenomenology of Spirit*. Translated by T. Pinkard. Cambridge: Cambridge University Press.

Heidegger, M. 1927/1962. Being and Time. Translated by J. Macquarrie & E. Robinson. Oxford: Basil Blackwell.

Heisenberg, W. 1925. Über quantentheoretische Umdeutung kinematischer und mechanischer Beziehungen. *Zeitschrift für Physik* 33: 879–893.

Heisenberg, W. 1958. *Physics and Philosophy: The Revolution in Modern Science*. New York: Penguin.

Heldke, L. 2019. Dewey and pragmatist feminist philosophy. In S. Fesmire (Ed.), *The Oxford Handbook of Dewey* (pp. 271–290). Oxford: Oxford University Press.

Helmholtz, H. 1873. *Popular Lectures in Scientific Subjects*. Translated by E. Atkinson (with introduction by J. Tyndale). London: Longmans Green.

Hempel, C. 1965. *Aspects of Scientific Explanation*. New York: Free Press.

Herbert, J. 1993. Peptides in the limbic system: Neurochemical codes for coordinated adaptive responses to behavioral and physiological demand. *Progress in Neurobiology* 41 (6): 723–791.

Herbert, J., & Schulkin, J. 2002. Neurochemical coding of adaptive responses in the limbic system. In D. Pfaff, A. P. Arnold, A. M. Etgen, S. E. Fahrbach, & R. T. Rubin (Eds.), *Hormones, Brain and Behavior* (pp. 659–689). New York: Elsevier Press.

Herrick, C. J. 1905. The central gustatory pathway in the brain of body fishes. *Journal of Comparative Neurology* 15 (5): 375–486.

Herrick, C. J. 1929. Anatomical patterns and behavioral patterns. *Physiological Zoology* 4: 439–448.

Herrick, C. J. 1948. *The Brain of the Tiger Salamander*. Chicago: University of Chicago Press.

Hickman, L. 2001. *Philosophical Tools for a Technological Culture: Putting Pragmatism to Work*. Bloomington: Indiana University Press.

Hickman, L. 2019. Dewey, pragmatism, technology. In S. Fesmire (Ed.), *The Oxford Handbook of Dewey* (pp. 492–506). Oxford: Oxford University Press.

Hildebrand, D. 2003. *Beyond Realism and Antirealism: John Dewey and the Neopragmatists*. Nashville, TN: Vanderbilt University Press.

Hildebrand, D. 2019. Dewey, Rorty, and Brandom: The challenges of linguistic pragmatism. In S. Fesmire (Ed.), *The Oxford Handbook of Dewey* (pp. 89–130). Oxford: Oxford University Press.

Hinde, R. 2002. *Why Good Is Good: The Sources of Morality*. London: Routledge.

Hohwy, J. 2013. *The Predictive Mind*. Oxford: Oxford University Press.

Hook, S. 1927. *The Metaphysics of Pragmatism*. Chicago: Open Court.

Hookway, C. 1984. Naturalism, fallibilism, and evolutionary epistemology. In C. Hookway (Ed.), *Minds, Machines, and Evolution*. Cambridge: Cambridge University Press.

Hubel, D. H., & Wiesel, T. N. 2005. *Brain and Visual Perception: The Story of a 25-Year Collaboration*. Oxford: Oxford University Press.

Huebner, D. (Ed.) 2018. *The Philosophy of Daniel Dennett.* Oxford: Oxford University Press.

Hume, D. 1739. *A Treatise of Human Nature.* Edited by L. A. Selby-Bigge. Oxford: Clarendon Press.

Hume, D. 1757/1965. *Treatise of Human Nature.* Edited by L. A. Selby-Bigge. Oxford: Clarendon Press.

Jackson, J. H. 1884/1958. Evolution and dissolution of the nervous system. In: J. Taylor (ed.), *Selected Writings of John Hughlings Jackson* (Vol. 2). London: Staples Press.

James, W. 1890. *The Principles of Psychology* (2 vols.). New York: Dover.

Jeannerod, M. 1997. *The Cognitive Neuroscience of Action.* Oxford: Blackwell Publishers.

Jeannerod, M. 1999. To act or not to act: Perspectives on the representation of action. *Quarterly Journal of Experimental Psychology* 52: 1–29.

Johnson, M. 1978. Kant's unified theory of beauty. *Journal of Aesthetics and Art Criticism* 38 (2): 167–178.

Johnson, M. 1987. *The Body in the Mind: The Bodily Basis of Meaning, Imagination, and Reason.* Chicago: University of Chicago Press.

Johnson, M. 1993. *Moral Imagination: Implications of Cognitive Science for Ethics.* Chicago: University of Chicago Press.

Johnson, M. 2007. *The Meaning of the Body: Aesthetics of Human Understanding.* Chicago: University of Chicago Press.

Johnson, M. 2014a. *Morality for Humans: Ethical Understanding from the Perspective of Cognitive Science.* Chicago: University of Chicago Press.

Johnson, M. 2014b. Experiencing language: What's missing in linguistic pragmatism? *European Journal of Pragmatism and American Philosophy* 6 (2): 14–27.

Johnson, M. 2017. *Embodied Mind, Meaning, and Reason: How Our Bodies Give Rise to Understanding.* Chicago: University of Chicago Press.

Johnson, M. 2018. *Aesthetics of Meaning and Thought: The Bodily Roots of Philosophy, Science, Morality, and Art.* Chicago: University of Chicago Press.

Johnson, M., & Schulkin, J. 2020. The common currency of aesthetic sensibility. *Transactions of the C.S. Peirce Society* 56 (3): 326–348.

Johnson, M., & Tucker, D. 2021. *Out of the Cave: A Natural Philosophy of Mind and Knowing.* Cambridge, MA: MIT Press.

Kagan, J. 1984. *The Nature of the Child.* New York: Basic Books.

Kagan, J. 2002. *Surprise, Uncertainty and Mental Structure.* Cambridge, MA: Harvard University Press.

Kahneman, D. 2011. *Thinking Fast and Slow*. New York: Farrar, Straus & Giroux.

Kahneman, D., Slovic, P., & Tversky, A. (Eds.) 1982. *Judgment Under Uncertainty: Heuristics and Biases*. New York: Cambridge University Press.

Kant, I. 1781. *Critique of Pure Reason*. Translated by W. Pluhar. Indianapolis: Hackett Publishing.

Kant, I. 1785/1981. *The Grounding for the Metaphysics of Morals*. Translated by J. Ellington. Indianapolis: Hackett Publishing.

Kant, I. 1787/1993. *The Critique of Practical Reason*. Translated by L. W. Beck. New York: Macmillan.

Kant, I. 1790/1987. *Critique of Judgment*. Translated by W. Pluhar. Indianapolis: Hackett.

Kant, I. 1800/2004. *Lectures on Logic: The Cambridge Edition of the Works of Immanuel Kant*. Edited by J. M. Young. Cambridge: Cambridge University Press.

Kellog, F. R. 1992. Who owns pragmatism? *Journal of Speculative Philosophy* 6: 67–80.

Keverne, E. B., & Curley, J. P. 2004. Vasopressin, oxytocin and social behavior. *Current Opinion in Neurobiology* 14: 777–783.

Keverne, E. B., & Curley, J. P. 2008. Epigenetics, brain evolution, and behavior. *Neuroendocrinology* 29: 398–412.

Kitcher, P. 2011. *The Ethical Project*. Cambridge, MA: Harvard University Press.

Kitcher, P. 2012. *Preludes to Pragmatism: Toward a Reconstruction of Philosophy*. Oxford: Oxford University Press.

Kitcher, P. 2018. Deweyian naturalism. In M. Bagger (Ed.), *Pragmatism and Naturalism* (pp. 66–87). New York: Columbia University Press.

Koch, C. 2004. *The Quest for Consciousness: A Neurobiological Approach*. Englewood, CO: Roberts & Company.

Koch, C., & Reid, R. 2012. Neuroscience: Observations of the mind. *Nature* 483 (21): 397–398.

Konorski, J. 1967. *Integrative Activity of the Brain*. Chicago: University of Chicago Press.

Koob, G. E., & LeMoal, M. 2002. *The Neurobiology of Drug Addiction*. Cambridge: Cambridge University Press.

Koopman, C. 2009. *Pragmatism as Transition: Historicity and Hope in James, Dewey, and Rorty*. New York: Columbia University Press.

Koopman, C. 2013. *Genealogy as Critique*. Bloomington: Indiana University Press.

Kornblith, H. (Ed.) 1980. *Naturalizing Epistemology*. Cambridge, MA: MIT Press.

Kornblith, H. 2014. *A Naturalistic Epistemology*. Oxford: Oxford University Press.

Kosslyn, S. M. 1980. *Image and Mind*. Cambridge, MA: Harvard University Press.

Kosslyn, S. M. 1994. *Image and Brain*. Cambridge, MA: MIT Press.

Kuhn, T. 1962. *The Structure of Scientific Revolutions*. Chicago: University of Chicago Press.

Lakoff, G. 1987. *Women, Fire, and Dangerous Things: What Categories Reveal about the Mind*. Chicago: University of Chicago Press.

Lakoff, G., & Johnson, M. 1980. *Metaphors We Live By*. Chicago: University of Chicago Press.

Lakoff, G., & Johnson, M. 1999. *Philosophy in the Flesh: The Embodied Mind and Its Challenge to Western Thought*. New York: Basic Books.

Lakoff, G., & Narayanan, S. in press. *The Neural Mind*.

Lakoff, G., & Nunez, R. 2000. *Where Mathematics Comes From: How the Embodied Mind Brings Mathematics into Being*. New York: Basic Books.

Langacker, R. 1987–91. *Foundations of Cognitive Grammar* (2 vols.). Stanford, CA: Stanford University Press.

Langer, S. 1937. *Philosophy in a New Key*. Cambridge, MA: Harvard University Press.

Langer, S. 1953. *Feeling and Form: A Theory of Art*. New York: Charles Scribner's Sons.

Langer, S. 1972. *Mind: An Essay on Human Feeling*. Baltimore, MD: Johns Hopkins University Press.

Larhammar, D., Dreborg, S., Larsson, T. A., & Sundstrom, G. 2009. Early duplication of opiate receptor and peptide genes in vertebrate evolution. *Trends in Comparative Endocrinology and Neurobiology* 1163: 451–453.

Leddy, T. 2012. *The Extraordinary in the Ordinary: The Aesthetics of Everyday Life*. Peterborough, ON: Broadview Press.

LeDoux, J. 1996. *The Emotional Brain: The Mysterious Underpinnings of Emotional Life*. New York: Simon & Schuster.

LeDoux, J. 2015. *Anxious: Using the Brain to Understand and Treat Fear and Anxiety*. New York: Viking Press.

Lehrman, D. S. 1953. A critique of Konrad Lorenz's theory of instinctive behavior. *Quarterly Review of Biology* 28: 337–363.

Light, A. 1996. *Environmental Pragmatism*. New York: Routledge.

Light, A., & Smith, J. 2005. *The Aesthetics of Everyday Life*. New York: Columbia University Press.

Locke, A. 1924. The concept of race as applied to social culture. In L. Harris (Ed.), *The Philosophy of Alain Locke: Harlem Renaissance and Beyond* (pp. 188–199). Philadelphia: Temple University Press.

Loewenstein, G. F. 1996. Out of control: Visceral influences on behavior. *Organizational Behavior & Human Decision Processes* 65 (3): 272–292.

Lorenz, K. 1970. *Studies in Animal and Human Behaviour* (Vol. 1). Cambridge, MA: Harvard University Press.

MacIntyre, A. 1981. *After Virtue: A Study in Moral Theory*. Notre Dame, IN: University of Notre Dame Press.

Madzia, R. 2016. Mind, symbol and action-prediction: George H. Mead and the embodied roots of language. In R. Madzia & M. Jung (Eds.), *Pragmatism and Embodied Cognitive Science* (pp. 213–232). Berlin: De Gruyter.

Madzia, R., & Jung, M. (Eds.) 2016. *Pragmatism and Embodied Cognitive Science*. Berlin: Walter de Gruyter.

Maier, N. R. F., & Schneirla, T. C. 1935/1964. *Principles of Animal Psychology*. New York: Dover Press.

Margolis, J. 2002. *Reinventing Pragmatism: American Philosophy at the End of the Twentieth Century*. Ithaca, NY: Cornell University Press.

Marler, P. 1961. The logical analysis of animal communication. *Journal of Theoretical Biology* 1 (3): 295–317.

Marler, P., & Hamilton, W. J. 1966. *Mechanisms of Animal Behavior*. New York: Wiley.

Martin, A. 1998. Organization of semantic knowledge and the origin of words in the brain. *Origin of Diversification of Language* 24: 69–87.

Martin, A. 2007. The representation of object concepts in the brain. *Annual Review of Psychology* 58: 25–45.

Martin, A., Wiggs, C. L., Ungerleider, L. G., & Haxby, J. V. 1996. Neural correlates of category specific knowledge. *Nature* 379: 649–652.

Maturana, H., & Varela, F. 1972. *Autopoiesis and Cognition: The Realization of the Living*. Boston Studies in the Philosophy and History of Science. Dordrecht: Reidel.

McCauley, R. 2011. *Why Religion Is Natural and Science Is Not*. Oxford: Oxford University Press.

McCumber, J. 2001. *Time in the Ditch: American Philosophy and the McCarthy Era*. Evanston, IL: Northwestern University Press.

McEwen, B. S. 1998. Protective and damaging effects of stress mediators. *New England Journal of Medicine* 338: 171–179.

McEwen, B. S. 2007. Physiology and neurobiology of stress and adaptation: Central role of the brain. *Physiological Reviews* 87: 873–904.

McEwen, B. S., & Stellar, E. 1993. Stress and the individual: Mechanisms leading to disease. *Archives of Internal Medicine* 153 (18): 2093–3101.

McEwen, B. S., Gray, J. D., & Nasca, C. 2015. Recognizing resilience: Learning from the effects of stress on the brain. *Neurobiology of Stress* 1 (January): 1–11.

McGinn, C. 1989. Can we solve the mind–body problem. *Mind* 98 (391): 349–366.

McKenna, E. 2013. *Pets, People, and Pragmatism.* New York: Fordham University Press.

McKenna, E., & Pratt, S. 2015. *American Philosophy: From Wounded Knee to the Present.* London: Bloomsbury.

Mead, G. H. 1928/1972. *The Philosophy of the Act* (Vol. 3). Chicago: University of Chicago Press.

Mead, G. H. 1934. *Mind, Self, and Society: From the Standpoint of a Social Behaviorist.* Chicago: University of Chicago Press.

Medina, J. 2011. Pragmatic pluralism, multiculturalism, and the new Hispanic. In G. Pappas (Ed.), *Pragmatism in the Americas* (pp. 199–226). New York: Fordham University Press.

Meltzoff, A. N., & Moore, M. K. 1977. Imitation of facial and manual gestures by human neonates. *Science 198: 75–78.*

Merleau-Ponty, M. 1962. *The Phenomenology of Perception.* Translated by C. Smith. London: Routledge & Kegan Paul.

Merleau-Ponty, M. 1963. *The Structure of Behavior.* Translated by C. Smith. London: Routledge & Kegan Paul.

Meyer, L. B. 1956/1973. *Emotion and Meaning in Music.* Chicago: University of Chicago Press.

Meyer, L. B. 1973. *Explaining Music: Essays and Explorations.* Berkeley: University of California Press.

Mikhail, J. 2000. *Rawls' linguistic analogy: A study of the "generative grammar" model of moral theory described by John Rawls in "A theory of justice"* (Ph.D. dissertation). Cornell University.

Mikhail, J. 2007. Universal moral grammar: Theory, evidence and the future. *TRENDS in Cognitive Science* 11 (4): 143–152.

Miller, N. E. 1959. Liberalization of basic S-R concepts: Extensions to conflict behavior, motivation and social learning. In S. Koch (Ed.), *Psychology: A Study of a Science* (Vol. 2, pp. 196–292). New York: McGraw-Hill.

Milner, D., & Goodale, M. A. 1995. *The Visual Brain in Action*. Oxford: Oxford University Press.

Misak, C. 2016. *Cambridge Pragmatism*. Oxford: Oxford University Press.

Mithen, S. 1996. *The Prehistory of the Mind: A Search for the Origins of Art, Religion, and Science*. London: Thames & Hudson.

Moore-Ede, M. C. 1986. Physiology of the circadian timing system: Predictive versus reactive homeostasis. *American Journal of Physiology* 250: R737–R752.

Moore-Ede, M. C., Sulzman, F. M., & Fuller, C. A. 1992. *The Clocks That Time Us*. Cambridge, MA: Harvard University Press.

Moreno, J. D. 1995. *Deciding Together*. Oxford: Oxford University Press.

Moreno, J. D. 1999. Bioethics is a naturalism. In G. McGee (Ed.), *Pragmatic Bioethics* (pp. 3–16). Cambridge, MA: MIT Press.

Moreno, J. D., & Schulkin, J. 2020. *The Brain in Context*. New York: Columbia University Press.

Mrosovsky, N. 1990. *Rheostasis: The Physiology of Change*. New York: Oxford University Press.

Nagel, T. 1979. *Mortal Questions*. Cambridge: Cambridge University Press.

Neisser, U. 1994. Multiple systems: A new approach to cognitive theory. *European Journal of Cognitive Psychology* 6: 225–241.

Nelson, R. J., Demas, G. E., Klein, S. L., & Kriegsfeld, L. J. 2002. *Seasonal Patterns of Stress Immune Function and Disease*. Cambridge: Cambridge University Press.

Neville, R. C. 1974. *The Cosmology of Freedom*. New Haven, CT: Yale University Press.

Neville, R. C. 1992. *The High Road Around Modernism*. Albany, NY: SUNY Press.

Niebuhr, R. 1932. *Moral Man and Immoral Society: A Study of Ethics and Politics*. New York: Charles Scribner's Sons.

Nietzsche, F. 1878/1984. *Human, All Too Human*. Translated by M. Faber & S. Lehmann. Lincoln: University of Nebraska Press.

Nietzsche, F. 1895/1990. *The Anti-Christ*. Translated by E. Hollingsworth. New York: Penguin.

Noddings, N. 2019. Dewey and the quest for certainty in education. In S. Fesmire (Ed.), *The Oxford Handbook of Dewey* (pp. 335–350). Oxford: Oxford University Press.

Norgren, R. 1995. Gustatory system. In G. Paxinos (Ed.), *The Rat Nervous System* (2nd ed., pp. 751–771). San Diego: Academic Press.

Norton-Smith, T. 2010. *The Dance of Person and Place: One Interpretation of American Indian Philosophy*. Albany, NY: SUNY Press.

Nussbaum, M. 1986. *The Fragility of Goodness: Luck and Ethics in Greek Tragedy and Philosophy*. Cambridge: Cambridge University Press.

Orosco, J.-A. 2011. Pragmatism, Latino intercultural citizenship, and the transformation of American democracy. In G. Pappas (Ed.), *Philosophy in the Americas* (pp. 227–244). New York: Fordham University Press.

Outlaw, L. 1996. *On Race and Philosophy*. New York: Routledge.

Panksepp, J. 1998. *Affective Neuroscience: The Foundations of Human and Animal Emotions*. New York: Oxford University Press.

Pappas, G. 2008. *John Dewey's Ethics: Democracy as Experience*. Bloomington: Indiana University Press.

Pappas, G. (Ed.) 2011. *Philosophy in the Americas*. New York: Fordham University Press.

Pappas, G. 2016. Dewey's radical logic: The function of the qualitative in thinking. *Transactions of the C.S. Peirce Society* 52 (3): 435–468.

Pashler, H. E. 1998. *The Psychology of Attention*. Cambridge, MA: MIT Press.

Pashler, H. E., & Harris, C. R. 2012. Is the replicability crisis overblown? *Perspectives in Psychological Science* 7: 531–536.

Peirce, C. S. 1868. Questions concerning certain faculties claimed for man. *Journal of Speculative Philosophy* 2: 103–114.

Peirce, C. S. 1878. Deduction, induction and hypothesis. *Popular Science Monthly* 13: 470–482.

Peirce, C. S. 1898/1992. *Reasoning and the Logic of Things*. Edited by K. L. Ketner & H. Putnam. Cambridge, MA: Harvard University Press.

Peirce, C. S., & Jastrow, J. 1884. On small differences of sensation. *Memoirs of the National Academy of Sciences* 3: 75–83.

Penfield, W. 1959. The interpretative cortex. *Science* 129: 1719–1725.

Perret, D. I., & Emery, N. J. 1994. Understanding the intentions of others from visual signals: Neurophysiological evidence. *Cahiers de Psychologie Cognitive* 13 (5): 683–694.

Peters, A., McEwen, B. S., & Friston, K. 2017. Uncertainty and stress: Why it causes diseases and how it is mastered by the brain. *Progress in Neurobiology* 156: 164–188.

Phelps, E. A. 2006. Emotion and cognition: Insights from the study of the human amygdala. *Annual Review of Psychology* 57: 27–53.

Phelps, E. A., O'Connor, K. J., Cunningham, W. A., Funayama, E. S., Gatenby, J. C., Gore, J. C., & Banaji, M. R. 2000. Performance on indirect measures of race evaluation predicts amygdala activation. *Journal of Cognitive Neuroscience* 12 (5): 729–738.

Pollick, A., & de Waal, F. 2007. Ape gestures and language evolution. *Proceedings of the National Academy of Sciences* 104 (19): 8184–8189.

Popper, K. 1959. *The Logic of Scientific Discovery*. London: Routledge.

Posner, R. 1993. *The Problems of Jurisprudence*. Cambridge, MA: Harvard University Press.

Power, M., & Schulkin, J. 2008. Anticipatory physiological regulation in feeding biology: Cephalic phase responses. *Appetite* 50: 194–206.

Powley, T. L. 1977. The ventromedial hypothalamic syndrome, satiety and cephalic phase. *Psychological Review* 84 (1): 89–126.

Powley, T. L. 2000. Vagal input to the enteric nervous system. *Gut* 47: iv30–iv32.

Pratt, S. L. 2002. *Native Pragmatism*. Bloomington: Indiana University Press.

Premack, D. 1990. The infant's theory of self-propelled objects. *Cognition* 36: 1–16.

Premack, D., & Premack, A. J. 1995. Origins of human social competence. In M. S. Gazzaniga (Ed.), *The Cognitive Neurosciences* (pp. 205–218). Cambridge, MA: MIT Press.

Price, H. 2013. *Expressivism, Pragmatism, and Representationalism*. Cambridge: Cambridge University Press.

Prinz, J. J. 2004. *Gut Reactions: A Perceptual Theory of Emotion*. Oxford; New York: Oxford University Press.

Putnam, H. 1981. *Reason, Truth, and History*. Cambridge: Cambridge University Press.

Putnam, H. 1983. *Realism and Reason: Philosophical Papers* (Vol. 3). Cambridge: Cambridge University Press.

Putnam, H. 2000. *The Collapse of the Fact/Value Distinction*. Cambridge, MA: Harvard University Press.

Quine, W. V. O. 1951. Two dogmas of empiricism. *Philosophical Review* 60: 20–43.

Quine, W. V. O. 1953/1961. *From a Logical Point of View*. New York: Harper Torchbooks.

Quine, W. V. O. 1960. *Word and Object*. Cambridge, MA: MIT Press.

Quine, W. V. O. 1969. Epistemology naturalized. In W. V. Quine (Ed.), *Ontological Relativity and Other Essays* (pp. 69–90). New York: Columbia University Press.

Ramachandran, V., & Hubbard, E. 2001. Synaesthesia: A window into perception, thought, and language. *Journal of Consciousness Studies* 8: 3–34.

Rauschecker, J. P., & Scott, S. K. 2009. Maps and streams in the auditory cortex: Nonhuman primates illuminate human speech processing. *Nature Neuroscience* 12 (6): 718–724.

Rawls, J. 1971. *A Theory of Justice*. Cambridge, MA: Harvard University Press.

Rawls, J. 1980. Kantian constructivism in moral theory. *Journal of Philosophy* 77 (9): 515–572.

Rawls, J. 1985. Justice as fairness: Political not metaphysical. *Philosophy and Public Affairs* 14 (3): 223–251.

Redmond, I. 1982. The salt mining elephants of Mount Elgon. *Wildlife* 24 (8): 288–293.

Rescorla, R. A., & Wagner, A. R. 1972. A theory of Pavlovian conditioning: Variations in the effectiveness of reinforcement and non-reinforcement. In W. J. Baker & W. Prokasy (Eds.), *Classical Conditioning: Current Research and Theory* (pp. 64–99). New York: Appleton-Century-Crofts.

Richter, C. P. 1943. *Total Self-Regulatory Functions in Animals and Man*. New York: Harvey Lecture Series.

Richter, C. P. 1965/1979. *Biological Clocks in Medicine and Psychiatry*. Springfield, IL: C. C. Thomas.

Ricoeur, P. 1984. *Time and Narrative* (Vol. 1). Translated by K. McLaughlin & D. Pellauer. Chicago: University of Chicago Press.

Rockwell, W. T. 2005. *Neither Brain nor Ghost: A Nondualist Alternative to the Mind-brain Identity Theory*. Cambridge, MA: MIT Press.

Rolls, E. T. 1996. The orbitofrontal cortex. *Philosophical Transactions of the Royal Society of London* 351 (1346): 1433–1443.

Rolls, E. T. 2000. The orbitofrontal cortex and reward. *Cerebral Cortex* 10 (3): 284–294.

Rolls, E. T., & Treves, A. 1998. *Neural Networks and Brain Function*. New York: Oxford University Press.

Rorty, R. (Ed.) 1967. *The Linguistic Turn: Recent Essays in Philosophical Method*. Chicago: University of Chicago Press.

Rorty, R. 1979. *Philosophy and the Mirror of Nature*. Princeton, NJ: Princeton University Press.

Rorty, R. 1982. *Consequences of Pragmatism*. Minneapolis: University of Minnesota Press.

Rorty, R. 1989. *Contingency, Irony, and Solidarity*. Cambridge: Cambridge University Press.

Rorty, R. 1990. *Philosophy and Social Hope*. New York: Penguin.

Rosch, E. 1975. Cognitive representations of semantic categories. *Journal of Experimental Psychology: General* 104 (3): 192–233.

Rosen, J. B., & Schulkin, J. 1998. From normal fear to pathological anxiety. *Psychological Review* 105: 325–350.

Rosenthal, S. D., & Bourgeois, P. L. 1980. *Pragmatism and Phenomenology*. Amsterdam: Gruner.

Rozin, P. 1976. The evolution of intelligence and access to the cognitive unconscious. In J. Sprague & A. N. Epstein (Eds.), *Progress in Psychobiology and Physiological Psychology* (pp. 246–280). New York: Academic Press.

Rozin, P. 1998. Evolution and development of brains and cultures: Some basic principles and interactions. In M. S. Gazzaniga & J. S. Altman (Eds.), *Brain and Mind: Evolutionary Perspectives* (pp. 111–123). Strasbourg: Human Frontiers Science Program.

Sabini, J., & Silver, M. 1982. *The Moralities of Everyday Life*. Oxford: Oxford University Press.

Sagoff, M. 2000. *The Economy of the Earth*. Cambridge: Cambridge University Press.

Saito, Y. 2021. The aesthetics of the everyday. In E. Xalta (Ed.), *The Stanford Encyclopedia of Philosophy*. Palo Alto, CA: Metaphysics Research Laboratory, Stanford University.

Sandel, M. 1982. *Liberalism and the Limits of Justice*. Cambridge: Cambridge University Press.

Saper, C. B., & Loewy, A. 1980. Efferent connections of the parabrachial nucleus in the rat. *Brain Research* 197 (2): 291–317.

Saper, C. B., Lu, J., Chou, T. C., & Gooley, J. 2005. The hypothalamic integrator for circadian rhythms. *Trends in Neuroscience* 28 (3): 152–157.

Sapolsky, R. 2018. *Behave: The Biology of Humans at Our Best and Worst*. New York: Penguin Random House.

Scanlon, T. M. 1998. *What We Owe Each Other*. Cambridge, MA: Harvard University Press.

Schacter, D. L. 1996. *Searching for Memory*. New York: Basic Books.

Schacter, D. L. 2019. Implicit memory, constructing memory and imagining the future: A career perspective. *Perspectives in Psychological Science* 14 (2): 256–272.

Schopenhauer, A. 1851/1973. *Essays and Aphorisms*. Translated by R. J. Hollingdale. New York: Penguin.

Schulkin, J. 1991. *Sodium Hunger*. Cambridge: Cambridge University Press.

Schulkin, J. 1999. *The Neuroendocrine Regulation of Behavior*. Cambridge: Cambridge University Press.

Schulkin, J. 2000. *Roots of Social Sensibility and Neural Function*. Cambridge, MA: MIT Press.

Schulkin, J. 2003. *Rethinking Homeostasis*. Cambridge, MA: MIT Press.

Schulkin, J. 2004. *Bodily Sensibility: Intelligent Action*. Oxford: Oxford University Press.

Schulkin, J. 2007. *Effort: A Neurobiological Perspective on the Will*. Mahwah, NJ: Erlbaum.

Schulkin, J. 2009. *Cognitive Adaptation: A Pragmatist Perspective*. Cambridge: Cambridge University Press.

Schulkin, J. 2011. *Adaptation and Well-Being: Social Allostasis*. Cambridge, MA: MIT Press.

Schulkin, J. (Ed.) 2012. *Action, Perception and the Brain: Adaptation and Cephalic Expression*. New York: Palgrave Macmillan.

Schulkin, J. 2013. *Reflections on the Musical Mind*. Princeton, NJ: Princeton University Press.

Schulkin, J. 2015. *Pragmatism and the Search for Coherence in Neuroscience*. London: Palgrave Macmillan.

Schulkin, J. 2017. *The CRF Signal: Uncovering an Information Molecule*. Oxford: Oxford University Press.

Schulkin, J., & Sterling, P. 2019. Allostasis: A brain-centered, predictive mode of physiological regulation. *Trends in Neurosciences* 42 (10): 740–752.

Schultz, W. 2002. Getting formal with dopamine and reward. *Neuron* 36 (2): 241–263.

Schultz, W. 2007. Multiple dopamine functions at different time courses. *Annual Review of Neuroscience* 30: 59–88.

Schultz, W., Preuschoff, K., & Camerer, C. 2008. Explicit neural signals reflecting reward uncertainty. *Philosophical Transactions of the Royal Society of London* 3: 3801–3811.

Schutz, A. 1932/1967. *The Phenomenology of the Social World*. Translated by G. Walsh & F. Lehnert. Evanston, IL: Northwestern University Press.

Searle, J. 1969. *Speech Acts: An Essay in the Philosophy of Language*. Cambridge: Cambridge University Press.

Searle, J. 1997. *The Mystery of Consciousness*. New York: New York Review of Books.

Seigfried, C. H. 1996. *Pragmatism and Feminism: Reweaving the Social Fabric*. Chicago: University of Chicago Press.

Sellars, W. 1956/1997. *Empiricism and the Philosophy of Mind*. Edited by R. Brandom. Cambridge, MA: Harvard University Press.

Sellars, W. 1962. *Science, Perception, and Reality*. New York: Routledge.

Sellars, W. 1968. *Science and Metaphysics*. New York: Humanities Press.

Seth, A. K. 2013. Interoceptive inference, emotion and the embodied self. *Trends in Cognitive Science* 17 (11): 565–573.

Shapin, S. 1995. *A Social History of Truth*. Chicago: University of Chicago Press.

Shapin, S., & Schaffer, S. 1985. *Leviathan and the Air-Pump*. Princeton, NJ: Princeton University Press.

Sheets-Johnstone, M. 1999. *The Primacy of Movement*. Amsterdam: John Benjamins.

Sherrington, C. 1906/1948. *The Integrative Action of the Nervous System*. New Haven, CT: Yale University Press.

Shettleworth, S. J. 1998. *Cognition, Evolution and Behavior*. Oxford: Oxford University Press.

Shook, J. R. 2000. *Dewey's Empirical Theory of Knowledge and Reality*. Nashville, TN: Vanderbilt University Press.

Shusterman, R. 2000. *Pragmatist Aesthetics: Living Beauty, Rethinking Art*. New York: Rowman & Littlefield.

Shusterman, R. 2008. *Body Consciousness: A Philosophy of Mindfulness and Somaesthetics*. Cambridge: Cambridge University Press.

Simon, H. A. 1982. *Models of Bounded Rationality*. Cambridge, MA: MIT Press.

Singer, P. 1981. *The Expanding Circle: Ethics and Sociobiology*. New York: Farrer, Straus, & Giroux.

Sinnott-Armstrong, W. (Ed.) 2008. *Moral Psychology* (3 vols.). Cambridge, MA: MIT Press.

Sleeper, R. W. 1986. *The Necessity of Pragmatism: John Dewey's Conception of Philosophy*. New Haven, CT: Yale University Press.

Smith, G. P. 1997. *Satiation from Gut to Brain*. Oxford: Oxford University Press.

Smith, J. E. 1961. *Reason and God: Encounters of Philosophy with Religion*. New Haven, CT: Yale University Press.

Smith, W. J. 1977. *The Behavior of Communicating: An Ethological Approach*. Cambridge, MA: Harvard University Press.

Solymosi, T., & Shook, J. R. (Eds.) 2014. *Neuroscience, Neurophilosophy, and Pragmatism: Brains at Work with the World*. New York: Palgrave Macmillan.

Sperry, R. W. 1961. Cerebral organization and behavior. *Science* 133 (3466): 1749–1757.

Sporns, O. 2011. *Networks of the Brain*. Cambridge, MA: MIT Press.

Starling, E. H. 1923. *The Wisdom of the Body: The Harveian Oration*. London: H. K. Lewis.

Steiner, P. 2014. Embodied cognitive science, pragmatism, and the fate of mental representation. In T. Solymosi & J. R. Shook (Eds.), *Pragmatism and Embodied Cognitive Science* (pp. 73–98). Berlin: Walter de Gruyter.

Stellar, E. 1954. The physiology of motivation. *Psychological Review* 61 (6): 5–22.

Stellar, J. R., & Stellar, E. 1985. *The Neurobiology of Motivation and Reward*. New York: Springer Verlag.

Sterling, P. 2004. Principles of allostasis. In J. Schulkin (Ed.), *Allostasis, Homeostasis and the Costs of Physiological Adaptation* (pp. 17–64). Cambridge: Cambridge University Press.

Sterling, P. 2020. *What Is Health? Allostasis and the Evolution of Human Design*. Cambridge, MA: MIT Press.

Sterling, P., & Eyer, J. 1988. Allostasis: A new paradigm to explain arousal pathology. In S. Fisher & J. Reason (Eds.), *Handbook of Life Stress, Cognition and Health* (pp. 629–649). Hoboken, NJ: John Wiley & Sons.

Sterling, P., & Laughlin, S. 2015. *Principles of Neural Design*. Cambridge, MA: MIT Press.

Stern, D. 1985. *The Interpersonal World of the Infant: A View from Psychoanalysis and Developmental Psychology*. New York: Basic Books.

Strand, F. L. 1999. *Neuropeptides: Regulators of Physiological Processes*. Cambridge, MA: MIT Press.

Stuhr, J. 1997. *Genealogical Pragmatism: Philosophy, Experience, and Community*. Albany, NY: SUNY Press.

Stuhr, J. 2003. *Experience and Criticism: John Dewey's Reconstruction in Philosophy*. Nashville, TN: Vanderbilt University Press.

Stuhr, J. 2016. *Pragmatic Fashions*. Bloomington: Indiana University Press.

Sullivan, S. 2001. *Living across and through Skins: Transactional Bodies, Pragmatism, and Feminism*. Bloomington: Indiana University Press.

Swanson, L. W. 2003a. The amygdala and its place in the cerebral hemisphere. *Annals of the New York Academy of Sciences* 985 (1): 174–184.

Swanson, L. W. 2003b. *Brain Architecture.* Oxford: Oxford University Press.

Taghert, P. H., & Nitabach, M. N. 2012. Peptide neuromodulation in invertebrate model system. *Neuron* 76 (1): 82–97.

Talisse, R. B. 2009. *Democracy and Moral Conflict.* Cambridge: Cambridge University Press.

Taylor, R. 1963. *Metaphysics.* Hillsboro, NJ: Prentice-Hall.

Testa, I., & Caruana, F. (Eds.) 2020. *Habits: Pragmatist Approaches from Cognitive Science, Neuroscience, and Social Theory.* Cambridge: Cambridge University Press.

Thompson, E. 2007. *Mind and Life: Biology, Phenomenology, and the Sciences of Mind.* Cambridge, MA: Harvard University Press.

Thompson, P., & Hilde, T. (Eds.) 2000. *The Agrarian Roots of Pragmatism.* Nashville, TN: Vanderbilt University Press.

Tikka, P. 2008. *Enactive Cinema: Simulatorium Eisensteinese.* Jyvaskyla, Finland: University of Art and Design Helsinki.

Tillich, P. 1951–1963. *Systematic Theology* (3 vols.). Chicago: University of Chicago Press.

Tinbergen, N. 1951. *The Study of Instinct.* New York: Oxford University Press.

Tinbergen, N. 1963. On aims and methods of ethology. *Zeitschrift Für Tierpsychologie* 20: 410–433.

Todd, P. M., & Gigerenzer, G. 2007. Environments that make us smart. *Current Directions in Psychological Science* 16 (3): 167–171.

Tolstoy, L. 1898/1995. What is art? In A. Neill & A. Ridley (Eds.), *The Philosophy of Art: Readings Ancient and Modern* (pp. 506–521). New York: McGraw-Hill.

Tomasello, M. 1999. *The Cultural Origins of Human Cognition.* Cambridge, MA: Harvard University Press.

Tomasello, M. 2014. *A Natural History of Human Thinking.* Cambridge, MA: Harvard University Press.

Tononi, G., & Koch, C. 2008. The neural correlates of consciousness: An update. *Annals of the New York Academy of Science* 1124 (1): 239–261.

Trevarthen, C. 1979. Communication and cooperation in early infancy: A description of primary intersubjectivity. In M. Bullowa (Ed.), *Before Speech* (pp. 321–347). Cambridge: Cambridge University Press.

Tucker, D. 2007. *Mind from Body: Experience from Neural Structure.* Oxford: Oxford University Press.

Tucker, D. M., & Luu, P. 2012. *Cognition and Neural Development*. Oxford: Oxford University Press.

Ungerleider, L., & Miskin, M. 1982. Two cortical visual systems. In D. Ingle, M. Goodale, & R. Mansfield (Eds.), *Analysis of Visual Behavior* (pp. 549–586). Cambridge, MA: MIT Press.

Valenstein, E. S. 2006. *The War of the Soups and the Sparks*. New York: Columbia University Press.

Varela, F. J., Thompson, E., & Rosch, E. 1991. *The Embodied Mind*. Cognitive Science & Human Experience Series. Cambridge, MA: MIT Press.

Von Frisch, K. 1953. *The Dancing Bees*. New York: Harcourt Brace Jovanovich.

Von Holst, E. 1973. *The Behavioral Physiology of Animals and Man* (Vol. 1). Coral Gables, FL: University of Miami Press.

Von Uexkull, J. 1957/2010. *A Foray into the Worlds of Animals and Humans: With a Theory of Meaning*. Translated by Joseph D. O'Neil. Minneapolis; London: University of Minnesota Press.

Weiss, P. A. 1939. *Principles of Development*. New York: Henry Holt.

Weissman, D. 1993. *Truth's Debt to Value*. New Haven, CT: Yale University Press.

Weitz, M. 1956. The role of theory in aesthetics. *Journal of Aesthetics and Art Criticism* 15 (1): 27–35.

West, C. 1989. *The American Evasion of Philosophy*. Madison: University of Wisconsin Press.

White, M. 1947. *Social Thought in America: The Revolt against Formalism*. Boston: Beacon Press.

Whitehead, A. N. 1929a. *Process and Reality: An Essay in Cosmology*. New York: Macmillan.

Whitehead, A. N. 1929b/1958. *The Function of Reason*. Boston: Beacon Press.

Whitt, L. 2009. *Science, Colonialism, and Indigenous Peoples: The Cultural Politics of Law and Knowledge*. Cambridge: Cambridge University Press.

Wiener, P. P. 1949. *Evolution and the Founders of Pragmatism*. Cambridge, MA: Harvard University Press.

Wilshire, B. 2000. *The Primal Roots of American Pragmatism*. University Park: Pennsylvania State University Press.

Wingfield, J. C. 2004. Allostatic load and life cycles: Implication for neuroendocrine control mechanisms. In J. Schulkin (Ed.), *Allostasis, Homeostasis and the Costs of Physiological Adaptation* (pp. 302–342). Cambridge: Cambridge University Press.

Winter, S. 2001. *A Clearing in the Forest: Law, Life, and Mind*. Chicago: University of Chicago Press.

Winter, S. L. 2021. "Who" or "what" is the rule of law? *Philosophy and Social Criticism* 47: 1–19.

Wittgenstein, L. 1953. *Philosophical Investigations*. Edited by G. E. M. Anscombe & R. Rhys. Translated by G. E. M. Anscombe. Oxford: Basil Blackwell.

Wolf, G. 1984. The place of the brain in an ocean of feelings. In J. B. Cobb Jr. & F. I. Gamwell (Eds.), *Existence and Actuality; Conversations with Charles Hartshorne* (pp. 167–184). Chicago: University of Chicago Press.

Wright, C. 1877. The evolution of self-consciousness. In F. Ryan (Ed.), *Chauncey Wright, Philosophical Discussions* (pp. 199–266). New York: Henry Holt.

Wundt, W. 1874/1893. *Principles of Physiological Psychology*. Translated by E. B. Titchener. London: Allen.

Young, P. T. 1959. The role of affective processes in learning and memory. *Psychological Review* 66 (2): 104–125.

Zeki, S. 1999. *Inner Vision: An Exploration of Art and the Brain*. Oxford: Oxford University Press.

Index

www.ingramcontent.com/pod-product-compliance
Lightning Source LLC
LaVergne TN
LVHW050928210926
848482LV00013B/570
9780262545167